Stephen Dando-Collins is the author
Australian history *Captain Bligh's Othe*
the Military Coup that Turned Australia in
published by Random House Australia.

He is also the author of a successful series of Roman histories published around the world by Wiley, with the most recent being the historical whodunit, *Blood of the Caesars*. His American histories, *Standing Bear is a Person* and *Tycoon's War*, are published in the US, UK and Canada by DaCapo Press, while his historical novel *The Inquest* has been published in the US, Canada, Spain and Italy. Stephen lives in Tasmania's Tamar Valley.

PASTEUR'S GAMBIT

LOUIS PASTEUR, THE AUSTRALASIAN RABBIT PLAGUE & A TEN MILLION DOLLAR PRIZE

STEPHEN DANDO-COLLINS

VINTAGE BOOKS
Australia

A Vintage book
Published by Random House Australia Pty Ltd
Level 3, 100 Pacific Highway, North Sydney, NSW 2060
www.randomhouse.com.au

First published by Vintage in 2008

Addresses for companies within the Random House Group can be found at
www.randomhouse.com.au/offices

National Library of Australia
Cataloguing-in-Publication Entry

Dando-Collins, Stephen.
Pasteur's gambit.

ISBN: 978 1 74166 703 5 (pbk.)

Loir, Adrien.
Pasteur, Louis, 1822-1895.
Pasteur Institute of Australia.
Rabbits–Control–Australia.
Livestock–Diseases–Australia.
Bacteriologists—France.

579.30994

Text designed and typeset by Midland Typesetters, Australia
Printed and bound by Griffin Press

Random House Australia uses papers that are natural, renewable and recyclable products and made from wood grown in sustainable forests. The logging and manufacturing processes are expected to conform to the environmental regulations of the country of origin.

CONTENTS

LIST OF ILLUSTRATIONS

ACKNOWLEDGMENTS

When I sat down to write this book, I felt a little like the mosquito that arrived at Bondi Beach to find it covered with sunbathers: I knew what to do, but didn't know where to begin.

The desk and floor of my red-walled study had become covered with the results of years of research – piles of books, documents, government reports, letters, telegrams, photocopies from private journals and scrapbooks, photographs, line drawings, cartoons, and hundreds of newspaper cuttings. The research material overflowed into the dining room, and threatened to take over the entire house.

Yet, it didn't take long for the crucial elements of this fascinating story to emerge. Here were a billion rabbits taking over a continent. Here was Louis Pasteur, and his great rival Robert Koch, fighting each other on Australia's shores. Here was Pasteur's dread of losing his Pasteur Institute before its doors had even opened. Here was his potential salvation in the Australian rabbit contest prize. And here was Pasteur's young nephew Adrien Loir, who would valiantly fight his uncle's battles in Australia, disobeying Pasteur time and again as his intuition served him well, to ultimately contribute millions of francs to Pasteur Institute funds from Australian sources.

By the time I had completed the book, I had come to understand Pasteur's fixations. For all his errors of judgment, and his bitter but understandable dislike of Germans, his crown as a great man who advanced science is still secure, as far as I am concerned. To my mind, Oliver Wendell Holmes' 1887 appraisal of Pasteur, which he wrote in his book *One Hundred Days in Europe* after meeting Pasteur in Paris, still stands: 'I look upon him as one of the greatest experimenters who ever lived, one of the truest benefactors of his race.'

By the time I had completed this book, I had also come to know, respect and very much like Adrien Loir, who is really the hero of this story. And then there is Sir Henry Parkes – what a colourful, likeable Australian character and prescient politician he was. And other Australians who became Adrien Loir's firm supporters, men such as Tiger Bruce, Arthur Devlin, and Walter and John de Villiers Lamb.

One or two Australians don't come out of this story so well – members of the Rabbit Commission and a few politicians in particular. Yet selfishness, short-sightedness, and professional jealousy are not qualities exclusive to Australians. Down through history, it has invariably been those who have neither the capacity nor the courage to take an original step who are the first to criticise those who do. In the end, Adrien Loir's persistence and the good intentions of many good Australians outweighed the damage done by those who opposed Pasteur in Australia.

Many kind and generous people helped me with various aspects of my research for this book, and I owe them my heartfelt gratitude. I most especially wish to thank Her Excellency the Governor of New South Wales, Professor Marie Bashir, for personally showing me her office, which was Sir Henry Parkes' office, in the old Chief Secretary's Department in Sydney; for allowing me to sit at her desk, which was Sir Henry's desk; for having the old Executive Council Chamber opened up for me to view, and so permit me to visualise the meetings of the Rabbit Commission which took place there and which I describe in this book; and for conducting my wife and myself around Government House, Sydney, in the footsteps of a number of the people who feature in this story, including Sir Henry Parkes, Lord Carrington and Adrien Loir.

My sincere thanks for her ongoing advice and research relating to my questions about Louis Pasteur, the Pasteur Institute, and senior institute figures go to Elisabeth Liber, Archivist of the Pasteur Museum, Paris. Thank you to the helpful Thiery Gasco, Cellar Master with Pommery Champagne at Reims, and his assistant Nathalie Dufresne. And special thanks to Madame Françoise Michel-Loir, granddaughter of Adrien Loir, in France, for providing me with extensive information about her grandparents in both letter and book form.

There are numerous scientific figures who also deserve my thanks for their wonderful cooperation, assistance and advice: Dr Tony Peacock, Chief Executive Officer of the University of Canberra's Invasive Animals Cooperative Research Centre; Jean-Claude Herrimans, Secretary of the Linnean Society of New South Wales, Sydney; Dr Joc Forsythe, former head of the Microbiological Unit and Public Health Laboratory at Melbourne University; Alda Theodora Scholtes of Melbourne University's School of Medicine; and Dr Jak Kelly of Sydney University and Vice President of the Royal Society of New South Wales.

And then there are the people who helped me unlock the secrets of the past in libraries and archives across Australia: Rosanne Walker, Librarian with the Basser Library, and Leah Moncrieff, Executive Assistant, Australian Academy of Science, Canberra; Rachel Pryor, Reference Librarian with the National Library of Australia, Canberra, and many NLA staff who bent over backwards to be of assistance both during the time my wife Louise and I spent researching at the library, and also in advance of our visit; Julie Wood and many other helpful staff at the Mitchell Library and the Reference Library, State Library of New South Wales, Sydney; Robyn Gurney, Archivist with State Records of NSW, Sydney; Grace Hui and the staff at the State Records of NSW Kingswood repository; staff of the State Library of Victoria, Melbourne; and staff of the State Library of Tasmania, Hobart.

Thank you to the astute Meredith Curnow, publisher at Random House Australia, for her ongoing encouragement and advice. Special thanks to my translation collaborators Traude Plays and the late Pierre

Plays, and Julie Logie of the Alliance Française, Tasmania. Thank you, too, to those who provided logistical help to me and Louise on our research expeditions, especially Rosemary and Malcolm Oliver and Mary and Andrew Fisher-Hyde.

I extend my everlasting gratitude to my New York literary agent, Richard Curtis. Both he and his wife and partner Leslie saw the potential of this book from the very beginning, and encouraged me at every step of the way. Richard has been my mentor for a decade now, pointing me in the right direction with my Roman, American, British, and Australian histories, and now also on the subject of Franco-Australian history.

And finally, my own honey bunny, my wife Louise, who shared my determination to find the wonderful human story lurking on the pages of all the piles of research material we jointly unearthed. If I am Mr History, as some call me, she is Mrs Remarkable.

AUTHOR'S NOTE

Conversion Rates

The conversion rate, from pounds sterling in the 1880s to present-day Australian dollars, I have taken from *Louis Pasteur and the Pasteur Institute in Australia* (Chaussivert and Blackman, editors). That publication is now two decades old, so the conversion of the £25,000 prize to $10 million in today's money is a conservative figure.

The conversion of £25,000 to 625,000 francs in 1887 is based on Louis Pasteur's figure, as published in *Le Temps* in November 1887.

1

THE KILLING TEAM ARRIVES

WITH FUSSY STEAM-POWERED tugboats nosing against her olive green hull and nudging her, like sheepdogs, in the direction they wanted her to go, the 3918-ton Royal Mail Ship *Cuzco* eased into Port Melbourne's Queen's Wharf. It was Sunday 1 April 1888 and the *Cuzco*'s forty-two-day journey from the other side of the world was at an end.

The *Cuzco* was the newest and fastest of the Orient Line's fleet of Royal Mail steamers on the England–Australia run. Every two weeks, one of the line's ships departed Plymouth for Australia and another sailed from Sydney for England, with others en route in between. This April Sunday, the *Cuzco*'s master Captain John Nixen watched with satisfaction from his open bridge as stevedores on the wharf dragged in the mooring lines. Apart from intense heat experienced while traversing the Red Sea after leaving the Suez Canal, the *Cuzco*'s voyage had passed without any event of note. It may have been more eventful had Captain Nixen's passengers been aware that one of their travelling companions had stored the microbes of a death-dealing virus in two small, innocuous-looking wooden boxes in his First Class stateroom. Such information could potentially generate panic, cause a riot, and motivate a mutiny. But not even Captain Nixen was privy to the passenger's secret.

The passenger in question was down on the main deck with his 350 fellow travellers, watching the busy wharf scene as the ship berthed. Twenty-four-year-old Adrien Charles Marie Loir was of medium height, well built, round-faced, with a modest moustache. He was impeccably dressed in the latest Parisian style – a tailored three-piece suit, silk tie and tie-pin, and white spats on his highly polished shoes. Adrien Loir was the nephew of one of the most famous men in the world – Dr Louis Pasteur, the legendary French microbiologist who had proven that germs cause illness, who had pioneered the vaccination of animals and humans, and whose research had changed the face of both medicine and agriculture. In France, Louis Pasteur was already a legend. In the rest of the world his name would live on in his pasteurisation process, in which germs are killed by the application of intense heat.

Adrien Loir's uncle had sent the young man and two colleagues to Australia to collect a £25,000 award, worth $10 million in today's money. This was the prize in one of the most unusual and lucrative international contests staged to that time. The award would be for launching biological warfare on Australia's shores.

Over the past six years, forty percent of the continent of Australia and much of New Zealand had been ravaged by a plague of rabbits. Hundreds of millions of furry English bunnies had been eating rural districts bare, destroying crops as effectively as locusts, and consuming the feed of the sheep and cattle on which the Australian and New Zealand economies had been booming until recently. Stock were dying in their thousands, and farms in outlying Australian districts were in danger of becoming dustbowls.

Rabbits are not native to Australia or New Zealand. And, unlike Europe, they have few natural enemies in the Antipodes. So, ever since two-dozen rabbits had been released into the wild in Victoria in 1859, the rabbit population of Australia had exploded. By 1883, rabbits were such a problem that special legislation was introduced into the parliaments of several Australian colonies to counter them. Four years and many millions of government and private pounds later, with the rabbits defying every measure employed against them, including trapping, hunting and poisoning, a desperate government in New South Wales, the most populous of the Australian colonies, had decided to run an international competition. This contest, advertised in all the major newspapers of the world in the second half of 1887, called for a provable biological method of safely wiping out the rabbit populations of Australia and New Zealand. Louis Pasteur, convinced that he had that remedy, and that the prize money was his, had dispatched nephew Loir and his companions to Australia to prove it, and to claim the reward.

With Adrien Loir had come Dr François E. Germont and Dr Frank Hinds. Both were older than Loir, but only by a few years. And unlike Loir, who was merely a medical student at this time, both had their doctorates in medicine. Yet despite his lack of formal qualifications, there was no doubt in the minds of his colleagues that Loir was the leader of the team – he had been Louis Pasteur's personal assistant and protégé for the past six years. In fact, Loir barely knew Germont and Hinds, and they barely knew Louis Pasteur.

Germont had been a nodding acquaintance of Loir's for the past year while he worked as an assistant to Dr Jacques-Joseph Grancher, one of Pasteur's most senior collaborators. Grancher ran the Pasteur rabies clinic in Paris, across town from Pasteur's Rue d'Ulm laboratory, producing and injecting the rabies vaccine that had come to dominate Pasteur's life and work. Germont had assisted Grancher in the rabies facility, occasionally running errands to the Rue d'Ulm where Loir worked with Pasteur.

Pasteur had in fact initially chosen another assistant to accompany Loir to Australia – Dr Louis Momont, nephew of Pasteur's colleague and disciple Dr Emile Roux. But at the last minute Pasteur had replaced Momont with Germont. This decision had been made after Germont gave the master his French translation of an English medical book. Impressed by this, Pasteur had decided that Germont's English skills would come in handy in Australia. Both Germont and team leader Loir knew that Germont had managed his translation only with the help of an English–French dictionary – Germont could not conduct even a basic conversation in English. But Pasteur had decreed that Germont would go to Australia instead of Momont, and few in Louis Pasteur's circle, especially among the *preparateurs* like Loir and Germont, had the temerity to correct their exalted leader or, worse, to disagree with him. This leadership principle, that Pasteur's word was law, was the strength of his method of operation and, occasionally, its weakness.

Not that Pasteur was expecting all the translation work of the Pasteur Mission to Australia – as he was calling Loir's team – to fall on François Germont's shoulders. On the recommendation of Sir Daniel Cooper, New South Wales' Agent-General in London, Pasteur had appointed a young English MD to the mission as interpreter, believing him to be fluent in both English and French. This too proved to be an error. Dr Frank Hinds had come to Paris from London in early February and spent ten days observing Pasteur at work in the Rue d'Ulm laboratory. Adrien Loir had been away from the laboratory at the time, making last-minute preparations for the overseas expedition, and had little involvement with Hinds. It was only once the mission was on its way to Australia that Loir realised that Hinds could barely speak a word of French – his own English was better than Hinds' French, and that was almost non-existent. It seems that Hinds had joined the mission merely to obtain a free passage to Australia and to beef up his *curriculum vitae* by claiming that he had been 'trained' by and worked for the famed Louis Pasteur.

It was obvious to Loir how Frank Hinds could have hoodwinked his uncle. Loir had worked with Pasteur continuously for a number of years now, more intensely and more closely than any of the better-known doctors with whom Pasteur collaborated. Loir knew that the single-minded master would often go an entire day in the laboratory without uttering a single word to the assistants working around him, pacing up and down as he thought through one biological problem or another. So as not to break Pasteur's concentration, no one was permitted to speak to him unless he spoke to them first. Obviously, Frank Hinds had negotiated the ten days in Pasteur's presence by nodding a great deal and uttering the occasional *oui* and *non*, and perhaps a *parfait* or *je comprend*.

Hinds' deception now presented Loir with a major problem. Before he could collect the £25,000 prize for his uncle, he would have to demonstrate the effectiveness of Pasteur's rabbit plague remedy by replicating the Pasteur experiments in front of the judges of the Intercolonial Royal Commission on Rabbit Destruction whose role was to recommend the winning competition entry. And of course those judges would ask questions. If Loir could not understand or satisfactorily answer those questions in English, how could he hope to convince the Commission of the efficacy of Pasteur's rabbit eradication method? If the Pasteur Mission was to succeed, and if Loir was to take home the £25,000 which he knew his uncle was determined to claim, the young man had to overcome the language barrier.

For years past, Pasteur had posed Loir seemingly insurmountable problems, and the devoted young Adrien had never let his uncle down, even on one occasion putting his life on the line for Pasteur. So Loir had resolved that this language problem was not going to prevent him from fulfilling his uncle's wishes. With an embarrassed Frank Hinds avoiding him on board the *Cuzco* as the ship steamed toward Australia, Loir, displaying the resourcefulness that would mark his time on this mission, had befriended a French-speaking British passenger, a Captain Armstrong, and begged him for English lessons. So, as the shipboard weeks passed, Loir learnt as much English as he could from the affable Armstrong while strolling the ship's decks, at dinner, and over games of whist in the First Class lounge. By the time the ship docked in Melbourne, Loir could just make himself understood in English but was still having difficulty comprehending what was said to him in the foreign tongue.

This was not the only problem he faced. On 28 March, when the *Cuzco* had reached Adelaide, its first Australian port of call, Loir had learned that ten days earlier the parliament of New South Wales had passed a law banning the importation of any foreign microbe. This seemed absurd to Loir. The Government of New South Wales was the principal sponsor of the rabbit extermination competition. The Premier of New South Wales, Sir Henry Parkes, had personally invited Louis Pasteur to send his rabbit-destroying microbes out to Sydney, capital of New South Wales, and here was Pasteur's nephew on his way to Sydney with those very microbes. This new law didn't make sense. Only later would Loir learn that there had recently been a power struggle within the New South Wales cabinet, and Pasteur's microbes had fallen victim to it.

Details of the new law had been given to Loir, almost certainly by the French consul to South Australia, as soon as the *Cuzco* docked in Adelaide. By the time a French-speaking correspondent of the Melbourne *Age* newspaper interviewed Loir that same day, the young man was still thinking through what he should do.

'I am bringing out with me microbes for the purpose of inoculating rabbits,' Loir had explained to the *Age* reporter, who knew all about Pasteur's participation in the international competition. When asked what his plans were now, Loir had replied, 'I regret that I cannot state at present what my plans are as far as Sydney is concerned. I understand there has been considerable legislation on the subject since my departure from Europe, and until I know how that legislation will affect me, my plans will have to remain in abeyance.'[1]

Loir's immediate concern was that the authorities in Sydney would seize and destroy his precious microbes. But even though he had been caught off-guard by this news of the inexplicable anti-microbe legislation, he had tried to remain upbeat hoping, like the good scientist he was, to find a way around the problem. 'Of course we are prepared to give a practical demonstration of the efficacy of my uncle's plan before the commission,' he had assured the *Age* reporter in Adelaide. 'I do not think there will be any doubt then that the remedy fully complies with the [competition] conditions as issued by the New South Wales Government.'[2]

Now, as the *Cuzco* tied up in Melbourne, Loir was gazing down at the noisy, crowded wharf. Labourers and hawkers milled around scores of horse-drawn wagons and carriages. Smiling locals in their Sunday best were waving to shipboard relatives, new migrants who had sailed across the world in the *Cuzco*'s steerage class on government-assisted passages of £8 per married couple and £6 per individual. Loir took in none of this; his mind was elsewhere. How should he now proceed on his uncle's behalf? It was not as if he could pick up a telephone and call Paris for orders; international telephone calls were still many years in the future. It was up to the twenty-four-year-old to decide the fate of the Pasteur Mission.

Loir raised his gaze. The port was full of ships and the young Frenchman's eye was suddenly caught by the French tricolour flying from the jack of a large steamer of the French Messagéries Maritimes Line docked nearby. Loir had a idea. Finding his friend Captain Armstrong, Loir asked if he knew where the French ship was bound. Armstrong said that it and others like it regularly sailed a circuit between Marseilles, Australia, Noumea in New Caledonia and back to France. How long would it take to sail from Noumea to Sydney? Five days, Loir was told. Loir nodded thoughtfully. A plan began to form in his mind.

During the voyage, Captain Armstrong would have given Loir a thorough briefing on Australia and New Zealand. At this time the Australian continent was divided into six British colonies. Apart from Western Australia, which was still ruled from England, these colonies, while technically under the control of Britain's Secretary of State for the Colonies in London were, like New Zealand, self-governing in every respect apart from foreign policy, which was dictated by Britain. Each colony had its own government, even had its own small army and navy, and was totally responsible for its finances. The Australasian colonies were like independent countries and they frequently acted this way, vying with each other, disagreeing with each other, and only occasionally working together in a common cause such as the rabbit destruction competition. It would be another thirteen years before the Australian colonies federated to become the single sovereign nation of Australia. The colonies would then become states within the federation, retaining their own state governments in addition to the central federal government. They would also retain their rivalry of old.

Adrien Loir realised that a law passed in New South Wales did not apply in the neighbouring colony of Victoria, where Loir now found himself. If he were to go on to Sydney as planned, his microbes could be confiscated under the new law. Technically, he should be able to land the Pasteur microbes here in Melbourne without fear of losing them. But to be on the safe side, he now decided to put his backup plan in motion.

As the *Cuzco*'s gangway was being lowered, Loir located François Germont and Frank Hinds. The Pasteur Mission would be landing here in Melbourne, Loir told his colleagues, and not going on to Sydney with the *Cuzco* as originally planned. Loir would give one box of microbes to the captain of the Messagéries Maritimes ship he had spotted in port, with the request that he take it to Noumea in New Caledonia – French territory – and safely store it there. If anything were to happen to the contents of the second box, which Loir would retain, it would require only a five-day voyage from Noumea to Sydney before the team would be able to quickly access the backup supply of vital Pasteur microbes.

Under Loir's plan, he and Germont would stay in Melbourne while English-speaker Hinds took the overnight train to Sydney. Hinds' job would be to sort out the problems regarding the importation of the microbes so the Pasteur Mission could demonstrate before the Rabbit Commission in Sydney and win the prize for Pasteur. As dissatisfied as Loir was with Frank Hinds, he didn't think the Englishmen could fail to communicate that, in compliance with the invitation issued by Premier Parkes, Loir was seeking official permission to bring the Pasteur microbes to Sydney and demonstrate them before the rabbit contest's judges without fear of them being seized and destroyed.

Followed by his two colleagues, down the *Cuzco* gangway came the dapper young Frenchman, bearing his two little wooden boxes. To the casual observer they would have resembled large cigar boxes. Each contained an hermetically sealed glass jar filled with a beef broth in which floated the deadly microbes. Once on the dock, Loir proceeded to the French ship nearby. The French captain very quickly agreed to take one of the little boxes to Noumea – anything for the nephew of his revered countryman Louis Pasteur! *Vive la France!* Then, with their baggage loaded onto the back of a hired carriage, the trio ventured into the city of Melbourne, heading for the Athenaeum Club, a private gentlemen's club.

Once they had left behind row after row of dockside wool stores and cool storage sheds that housed the meat which since the 1870s had come to rival wool as Australia's major export, and passed the first of the many painted wall advertisements for Robur Tea that the new arrivals would encounter on Australia's shores, the streets of the sprawling city of half a million people suddenly became deserted. As Loir was to learn, Sunday in the Australian colonies was the day of rest – by law. No shop, office nor public building was open. Sporting grounds in the sports-crazy city were locked, theatres and taverns were closed. It was as if it were illegal to enjoy the day off. Loir was to discover that it was even against the law to hunt on Sundays. The only buildings open to the public were churches, hotels – with their bars closed – and private gentlemen's clubs.

Just fifty-three years earlier, as the first settlers from across the water in Van Diemen's Land had erected their wattle and daub huts beside the Yarra River, Melbourne had been nothing but bushland. Built on the back of a gold rush and the golden fleece, Melbourne was now Australasia's largest city. With its grey stone public edifices and block after block of hotels, shops and theatres, downtown Melbourne spread beside the narrow Yarra, which was not unlike Paris's River Seine. In fact, Melbourne liked to call itself 'the Paris of the South'.

The members of the Pasteur Mission arrived at the very posh Athenaeum Club in Collins Street. In all probability their escort was either Captain Armstrong or another of the *Cuzco*'s First Class passengers – the short-term honorary membership that Loir and company sought could only have come on the recommendation of an existing member. Quite possibly, the snooty officials of the Athenaeum would have wondered whether this was some sort of elaborate April Fool's Day prank, with the nephew of the world-famous Louis Pasteur applying for membership of their club on 1 April. But Loir was able to produce his credentials, and he and his two colleagues were granted one month's honorary membership of the Athenaeum, effective immediately, which allowed them to take rooms at the club and use its excellent dining room.

The Pasteur Mission to Australia had landed on the great southern continent, but they were still well-short of their destination, and a long way from winning the rabbit-destruction prize that Louis Pasteur had sent them to claim.

***

Next day, a Monday, young team leader Adrien Loir took Frank Hinds to Melbourne's Spencer Street Station and put him on the Sydney Express. Hinds' overnight rail journey to Sydney would include a stopover at Albury on the Victoria–New South Wales border to change trains – typically, the competing Australian colonies had built their railways to different gauges, making it impossible for the trains from one colony to run on the tracks of another.

Loir, who paid for Hinds' rail fare, was already worrying about money. At home his uncle always practised the strictest economy, living quite frugally, and Pasteur expected those around him to do the same. Pasteur had paid the First Class steamer fare – £70 each or $28,000 in today's money, from Plymouth to Sydney – to send Loir and his team members to Australia but he had given his nephew very little spending money, convinced that the team should be able to complete their task within six weeks at the most and then be on their way back to France. On the Tuesday, two days after arriving in Melbourne, Loir had also accepted an honorary membership of the less expensive Australian Club and, the next day, of the Melbourne Club where room and board were cheaper again.

He had already given Frank Hinds some of the funds he'd brought from France to cover the Englishman's expenses in Sydney and, not knowing how long he and Germont might have to wait in Melbourne before they could follow Hinds, Loir was determined to economise. Gaining temporary membership of these other Melbourne clubs was not a problem for Loir was welcomed wherever he went in Melbourne; as the nephew of the famous Louis Pasteur he was something of a celebrity. Besides, it turned out that despite his poor English, young Adrien Loir had the knack of charming just about everyone he met.

One of his new friends was to help Loir follow instructions from Louis Pasteur. Before Loir left Paris, Pasteur had told him that as soon as he landed in Australia he was to determine the virulence of his chicken-cholera microbes. There was a distinct possibility that the microbes, microscopic living organisms, had died in transit. Loir was particularly concerned that the extreme heat they'd been exposed to while traversing the Red Sea had killed them, for he knew that microbes could be adversely affected by high temperatures. Obviously, if the microbes were dead, the Australian mission was at an end.

To determine the virulence of the microbes, Loir needed test rabbits. As he soon learned, it was illegal for anyone not registered with the local Medical Board to conduct medical experiments. Any person assisting in the performance of such experiments was also liable to prosecution. And it was a crime, punishable by a fine of £100, to sell, breed or keep rabbits in Victoria.[3] Just the same, Loir's new gentleman friend, whose name Loir never revealed, happily risked a fine and even imprisonment to quickly secure two healthy rabbits for Loir's experiment.

To ensure that Loir and his helper were kept out of jail, secrecy was all important, so Loir conducted his experiment in the gentleman's Melbourne home, his cohort keeping the two rabbits in his wife's dressing room. What the lady thought about this is not recorded. Meanwhile, Loir had to prepare a culture for the test from his precious supply of beef broth. In the laboratory he would have done this using an incubator at a constant temperature of 37 degrees centigrade. Without access to an incubator, he had to find an alternative.

As always, the resourceful Adrien came up with a solution. He crafted himself a special belt. Then, unsealing the wooden box's glass jar, he poured a small quantity of the beef broth into a glass vial which went into the belt. For the next forty-eight hours he kept the belt around his waist, under his clothes, in the hope that his body temperature would serve the same purpose as an incubator. If the microbes had not lost their virulence, it would only require one drop of the lethal beef broth on the rabbits' feed for it to take effect. In repeated experiments in France by Loir and Pasteur, the microbes had killed test rabbits in under twenty-four hours. At first, the test animals would show no ill effects, but once the microbes did their job, the rabbits' hearts would suddenly stop.

No doubt nervously, and with his secret assistant watching in awe, Loir applied the culture to some lettuce and gave it to the test bunnies. Happily for Loir but unhappily for the rabbits, both animals died, swiftly and painlessly, before a day had passed. The microbes were as potent as the day that Louis Pasteur had personally supervised the filling of the glass flasks in his Paris laboratory back in February. Now Adrien Loir was ready to prove to the Australasian judging panel that the Pasteur biological method for rabbit destruction met all the specifications of the international competition, and take the prize home to Paris. As he must. For, as Loir was all too aware, there was more to this than professional pride or national glory, although both played a part. Indeed, more was riding on the prize than Loir or Pasteur would ever admit. Louis Pasteur was not just confident of winning the $10 million prize. Pasteur was desperate for the money.

For young Adrien Loir, the story of Louis Pasteur's bid to win the Australian prize – an incredible story that would combine science, sabotage, subterfuge, and even an interlude with the sexiest, most famous actress in the world – was about to begin in earnest.

2

PASTEUR'S BEER OF REVENGE

ONE OF ADRIEN LOIR'S earliest memories of his Uncle Louis harked back to the winter of 1870–71. Adrien was eight at the time. For weeks the boy had been wearing the stripes of a quartermaster sergeant on his fur-trimmed overcoat, sewn there by the daughter of a family friend in the National Guard militia. Adrien would march around the Loir family house at Lyon shooting imaginary enemies with an imaginary gun. France was at war. First the country had been invaded by the German states, led by Prussia, during the brief and, for France, humiliating Franco–Prussian War. German armies had occupied much of northeastern France and surrounded Paris in a matter of weeks, forcing the abdication of France's Emperor Napoleon III.

Now France was racked by internal troubles, with rebellious so-called communes set up in Paris and other cities, including Lyon in south-central France, defying the national government. To escape the Germans, Louis Pasteur, his wife Marie and their daughter Marie-Louise, had come to stay with their in-laws the Loirs at their house in the centre of Lyon. Adrien Loir Senior, young Adrien's father, had worked with Louis Pasteur at the University of Strasbourg in Alsace-Lorraine in the 1850s, sharing an admiration for the daughters of the university's rector, Charles Laurent. After a brief courtship, Pasteur had married the elder daughter, the tall, attractive Marie; while Dr Loir had married the younger, more highly strung Amelie. The sisters had remained close after the Pasteurs moved to Paris and the Loirs to Lyon, where the two men took up senior professorships – Pasteur with the École Normale Supérieure (ENS), the prestigious academy founded by Napoleon Bonaparte in Paris; and Dr Loir with the University of Lyon.

In Lyon, this particular winter's night that was to lodge so firmly in young Adrien's memory, the two families had just sat down to dinner and were sipping their soup when, nearby in the Place Louis XVI, a drummer of the National Guard began to beat 'To Arms'. The rebels of the Lyon Commune, the 'communards' as they were known, had just assassinated the Guard's Commander Arnaud. Little Adrien's father, an officer with the Guard, jumped up from the table and rushed upstairs to throw on his uniform. Meanwhile, Adrien's mother Amelie dutifully hurried to secure her husband's sword; when Dr Loir dashed back down from the bedroom, Amelie helped him strap it on. With a hasty kiss on the cheek for his anxious wife, Dr Loir then went to join his unit.

All this time Louis Pasteur had remained sitting, stone-faced, at the table, continuing to sip his soup. Young Adrien was dumfounded. 'To my mind,' he would write many years later, thinking back to that day, 'this was war. It seemed strange to me that Pasteur should not move.'[4] As Adrien would later come to realise, there was nothing his uncle could have done that day. Many months before, Louis had attempted to volunteer for the National Guard, only to be rejected because of the lingering effects of a stroke he had suffered in 1868 which had left his left hand partially paralysed. And as he was not a doctor of medicine – Pasteur's doctorate was in chemistry – he was not even recruited to help with the wounded. To his own great frustration Pasteur, the son of a sergeant in Napoleon Bonaparte's army who had been awarded the *Légion D'honneur* by Bonaparte personally for his bravery, could do no more than look after his family and try to keep them safe.

Before long, the French Government would bloodily put down

the communes, but republican government was cemented in France. Peace had a bitter price for the French; they had to agree to humbling terms with Prussia's 'Iron Chancellor', Otto von Bismarck, paying huge financial reparations and handing over much of Alsace and Lorraine, including the Rhine city of Strasbourg where the Pasteurs and the Loirs had been married. As Young Adrien Loir came to appreciate, Louis Pasteur never forgave the Germans. It was not only the national humiliation. On the eve of the war, Napoleon III had promised Pasteur a seat in the French Senate; the war that took away the emperor rendered this promise valueless.

Pasteur had declared that to his dying day all his future studies would carry the dedication, 'Hatred to Prussia. Vengeance! Vengeance!'[5] To register his disgust with the Germans, he had sent back an honorary doctorate given to him by the University of Bonn.

Once he was back at his laboratory at the ENS in Paris Pasteur went in search of a practical way of exacting revenge by helping the nation beat the Germans at their own game, through beer: the 'beer of revenge' he called it. 'Our [France's] misfortunes inspired me with the idea,' he was to say.[6]

At the time, the Germans were the acknowledged experts in brewing beer. They had been doing it since Roman times, and probably well before. France's best hop fields were in Alsace and Lorraine, which since the war had become German territory, and Pasteur's assistant Emile Duclaux had specialised in the study of brewer's yeast. It also helped that Pasteur had already done a great deal of research into the subject of fermentation.

There were two types of beer. The British used the high fermentation technique to produce the darker beers – porter, ale, pale ale, stout, bitter and alestout. In the 1860s German brewers in Bavaria had perfected the low fermentation technique, involving lower temperatures at the fermentation stage. This produced the so-called blond beer, and this style had spread to Prussia, Bohemia, Austria, and then to France. If high beers weren't drunk soon after brewing they went off, and hence did not travel well. Low beers had much better keeping qualities and could be stored for some time, making them ideal for export. France's low beer was still inferior to the

German product so Pasteur set himself the task of making it not only better than German beer but more economical to produce.

Pasteur had already attained fame by creating the process that bears his name – pasteurisation – whereby liquids such as milk, wine and beer are subjected to rapid heat treatment to kill germs. By August 1871, Pasteur had perfected a new process, which he patented, that enabled low fermentation in any climate without contamination. But France's brewers were concerned that Pasteur's process would destroy the unique flavours of their beers, and failed to adopt it.

Still, Pasteur's work was not in vain. It was to be Denmark, another victim of German imperialism, that was to profit most from Pasteur's beer research. In 1870, Karl Jacobsen opened a large new brewery in Copenhagen. The entrepreneurial Jacobsen family, makers of Carlsberg Beer, took Pasteur's new fermentation method on board. As a consequence of this decision, within thirteen years the Jacobsens would increase the worldwide annual export of Carlsberg Beer from four million hectolitres to 200 million hectolitres.[7] The Jacobsen family freely attributed their success to the science of Louis Pasteur, even erecting a bust of him at their brewery. In the end, to Pasteur's disappointment, his beer was not France's 'beer of revenge' but Denmark's.

Young Adrien Loir could not know it when Pasteur stayed with his family over that grim winter of 1870–71, but he would grow up to become his uncle's protégé, would one day go to Copenhagen to help the brewers of Carlsberg Beer, and would lead a team to Australia to claim a prize that Louis Pasteur desperately needed.

3

BECOMING PASTEUR'S PROTÉGÉ

WHEN, IN 1875, Marie Pasteur invited her sister Amelie Loir to spend the summer with the Pasteurs at Arbois in the Jura Mountains, Amelie gladly collected her only child, twelve-year-old Adrien, and took the train from Lyon to Arbois. Louis Pasteur had by this time turned the Arbois house of his late father, Jean-Baptiste Pasteur, a tanner by trade, into his summer holiday retreat. Pasteur used three rooms upstairs as bedrooms for himself, Marie and their guests, while Pasteur's sister and brother-in-law, who continued to operate the family tannery, lived downstairs. For Adrien Loir and his mother, the annual trek to Arbois for the summer became *de rigueur*, and something Adrien greatly looked forward to. A bright child, Adrien impressed his Uncle Louis and, to Pasteur's delight, the boy took a keen interest in his scientific work – unlike his own son Jean-Baptiste, who had pursued law and would join the French diplomatic corps. A firm bond between uncle and nephew began to grow.

In 1878, fifteen-year-old Adrien had to write a school project about a famous person of his choice. Not surprisingly, Adrien chose to write about his microbiologist uncle who had become France's most famous scientist. More than just lauding Pasteur, Adrien attempted to describe his uncle's career to show why he deserved to be considered a great man, giving examples of the problems he'd had to overcome and how he had gone about it. It so happened that Pasteur had been visiting the Loirs at Lyon as Adrien was completing his essay, and Adrien's parents asked their boy to read his draft paper to its subject. Pasteur listened in impassive silence as Adrien nervously read his essay aloud. Pasteur then took the essay from Adrien, corrected errors of fact with a pencil, then handed it back.

There is no record of how Adrien's teacher received the essay but Pasteur liked it so much that he kept a copy. Several years later, when his daughter Marie-Louise married journalist René Valéry-Radot, Pasteur passed young Adrien's paper onto him. Valéry-Radot subsequently used Adrien's essay as the basis for his book about his father-in-law, *Monsieur Pasteur: History of a Scientist by a Layman*,[8] published in 1883.

By 1880, Pasteur had made a decision about young Adrien. The scientist had retired from his ENS professorship eight years before but had been permitted to retain his laboratory on the Rue d'Ulm on Paris's Left Bank and was granted a small government pension. By the summer of 1880, Pasteur was fifty-seven and, having just buried his last remaining sister Virginie, was feeling his mortality and considering his legacy. Increasingly hampered by his weak left hand, he had to rely on his students to physically carry out his experiments to his instructions. Some of those students could not handle his authoritarian style and soon left; or, if they did last, it would often only be for a matter of months because they would receive a mandatory government teaching appointment and have to move on. Some of Pasteur's former assistants became his long-term collaborators but what he needed was the day-to-day mind and muscle of a dedicated permanent assistant under his roof and under his control – basically, a clever robot.

And then there was the matter of the master's legacy. To his mind, the future of the name of Louis Pasteur would be assured if a family member were to become his scientific protégé. With his own son pursuing the law, Pasteur saw young Adrien Loir as that protégé. In choosing Loir, he also had an eye to professional security.

As Loir himself was to later say, Pasteur knew that in taking on his nephew he could trust him to keep Pasteur's private views within the family.[9]

For his part Adrien had harboured no ambition to become a scientist. Once he achieved his *baccalauréat*, or bachelor's degree, at Lyon, he'd been planning to enter the renowned French military academy of Saint-Cyr to become a dashing army officer. But Loir's single-minded uncle was not interested in the youngster's dreams. Announcing to Loir's parents that their son was to become his personal assistant, without any discussion on the subject with either Adrien or his mother or father, Pasteur gave the Loirs a list of studies that he expected Adrien to complete over the next two years in preparation for joining the master at his Rue d'Ulm laboratory. Adrien's father, who was now Dean of the Faculty of Sciences at Lyon University, was instructed to thoroughly train his son in laboratory techniques. Also on Pasteur's orders, Adrien was to undertake courses in glass-blowing and calligraphy, both of which Pasteur considered handy for a chemist.

The Loirs were hugely flattered that Pasteur had chosen their son to work with him; after all, Pasteur was considered one of the most brilliant scientists in the world. Six years before, for example, Scottish surgeon Joseph Lister, the father of antiseptic medicine, wrote to Pasteur to thank him 'for having, through your brilliant researches, demonstrated to me the truth of the germ theory of putrefaction'.[10] The Loirs agreed to Pasteur's plans for Adrien without a hint of dissent, while Adrien himself had no say in the matter. Over the next two summers when Adrien joined the Pasteurs at Arbois, Pasteur talked at length with the young man about his work, and confided secrets that Loir would keep close for the next fifty-seven years.

But to go back to February 1880: Pasteur announced that, following eighteen months of research with the microbe that caused chicken-cholera, a deadly disease in fowls, he had developed a vaccine against the disease. Despite its name, chicken-cholera is unrelated to cholera and, unlike cholera, is harmless to man. Chicken-cholera outbreaks could sweep through farmyards killing fowls wholesale and destroying the livelihoods of French poultry farmers. So, in announcing that he had created a preventative via vaccination, Pasteur immediately grabbed the nation's attention.

The idea of vaccination against disease was not new. The principle of vaccination had first been promoted by English physician Edward Jenner who, in 1796, injected a weakened strain of cowpox into the bloodstream of an eight-year-old boy, correctly believing that it would cause the body to develop immunity to smallpox. While the benefits of vaccination against smallpox were slow to be accepted in Britain, Napoleon Bonaparte soon became a believer, inoculating the entire French army in 1805.

Now, seventy-five years later, Pasteur was one of the tiny minority of the world's scientists who believed that animals and humans could routinely be given immunity against a variety of diseases by inoculating them with a vaccine made from the microbes responsible for the disease. Most people at the time thought the idea of injecting a disease into the body to be insanely homicidal. Creating a vaccine, meanwhile, depended on first isolating the microbe that caused the disease in question. Subsequently, figuring out a way to create a vaccine based on that microbe was like finding a needle in a haystack. Just as every disease is different, so is the path to defeating each one. Via trial and error, Pasteur had found the path to a vaccine for chicken-cholera.

The breakthrough had its foundation two years earlier when Pasteur had been sent the head of a cockerel that had died from the disease by thirty-one-year-old Professor Jean-Joseph-Henri Toussaint of the Toulouse Veterinary School. From the blood of the dead bird, Pasteur had been able to cultivate chicken-cholera microbes from which he had produced a vaccine. But once he'd announced his breakthrough, Pasteur was unusually reticent about revealing how he had cultivated the attenuated, or weakened, disease microbes that were the basis of his chicken-cholera vaccine.

He had a very good professional if selfish reason for his reticence. Pasteur had been the first to appreciate that the air around us is full of germs. Just as Lister had applied Pasteur's germ theory to protecting open wounds from the air by using an antiseptic such as carbolic acid, Pasteur had realised that he could also reverse the germ theory to cultivate germs and create a vaccine. Basically, all he'd done to create his vaccine was expose a culture of the chicken-cholera microbe to the air in a weak acidic medium for an extended period of time. Pasteur was reluctant to share the secret of that very simple culturing process because he knew his rivals could apply it to other microbial diseases, and he wanted a running start on research in other areas.

Pasteur was particularly keen to produce a vaccine for anthrax, a deadly disease in sheep, cattle, oxen and horses that was known in the time of Moses. By the 1870s anthrax was killing hundreds of thousands of head of stock in France and across Europe. In Russia it was called Siberian plague. In Britain it was variously known as splenic fever because one symptom was a grossly enlarged spleen, as well as malignant pustule and woolsorters' disease. In France it was called *charbon*. In Germany it used to be called *Milzbrand*, but was referred to as anthrax, which became its universal name. This was primarily because it was a German, Dr Robert Koch, who made the initial breakthrough with anthrax and popularised the term.

The anthrax bacillus had first been identified in 1850 by French physicians Pierre Rayer and Casmir Davaine; German veterinarian Franz Pollender later claimed to have independently discovered it in 1849. When Robert Koch was merely a district medical officer at Wollstein in the Prussian province of Posnan, he had spent four years between 1872 and 1876 studying the anthrax bacillus in fine detail, even photographing it, and discovering its quite frightening nature. 'It was he who showed the role of the spore in the etiology of anthrax,' Pasteur's former assistant and later collaborator Emile Duclaux was to write of Koch, 'and he did it in a way truly marvellous for its simplicity.'[11]

Through experiments in his primitive home laboratory in the 1870s, Koch had been able to show that anthrax disease could be transmitted from one animal to another via inoculations of blood contaminated with *Bacillus anthracis*. He had then created pure cultures of the bacilli by growing them on the aqueous humour of an ox's eye. He had found that, in unfavourable conditions, anthrax bacilli form protective spores that resist such adverse conditions as lack of oxygen.

After his findings were published in August 1876, Koch had said no more about anthrax, not even when French physiologist Paul Bert disputed his findings five months later with (flawed) research of his own. And, most importantly, Koch had not attempted to proceed to the next stage and develop a vaccine against anthrax – because he did not believe in vaccination. Koch, who was more interested in causes than cures, had moved on to mapping other parts of the bacterial world.

As a result of an 1877 request from the French Department of Agriculture, Louis Pasteur had been using Koch's findings as the basis of a vaccine-focused study of anthrax, and by early 1880 he had high hopes of adapting his chicken-cholera culturing techniques to developing an anthrax vaccine – which Koch had refused to even contemplate. But by August 1880, just five months after Pasteur announced his chicken-cholera breakthrough, he was still a long way from developing an anthrax vaccine. It was at this point that young Professor Toussaint in Toulouse, the same Toussaint who had sent Pasteur the rooster's head for his chicken-cholera research, let it be known that *he* had developed a vaccine for anthrax based on *Bacillus anthracis*. But Toussaint would not say how he had done it.

When word of Toussaint's apparent discovery became known at the Academy of Medicine in Paris, Gabriel Colin, head of France's leading veterinary school at Alfort, declared out of the blue that he could prove anthrax was not caused by *Bacillus anthracis* but by another virulent agent. As Emile Duclaux was to say, 'Science is like a train . . . which, after having gone forward, sometimes goes backward.'[12] The Colin argument certainly set anthrax research in reverse, if only briefly.

When Toussaint failed to make details of his discovery public, both he and Pasteur were admonished in the Academy for announcing 'secret remedies' which they would not detail – in Pasteur's case they were referring to his chicken-cholera vaccine. The affronted Pasteur threatened to resign from the Academy which, because of his vaunted reputation, apologised to him. But the younger, less-renowned Toussaint was forced to reveal the secret of his anthrax vaccine before he had conducted exhaustive tests on it.

Pasteur was summering at Arbois when he learned the details of Toussaint's process. Toussaint had vaccinated his test animals with anthrax-affected blood heated to 55 degrees centigrade, which he claimed killed the anthrax bacilli. Pasteur was astonished. Toussaint's method killed the bacilli? Then how could the vaccine work? This didn't add up for Pasteur, whose basic principle of vaccination revolved around weakened but still very much alive microbes. 'Ten times a day I have thought about taking the train to Paris,' he wrote to a colleague. 'I will not believe this astounding fact until I have seen it with my own eyes.'[13]

He promptly ordered his new assistant Dr Charles Chamberland to Toulouse to reproduce Toussaint's experiments. When Chamberland did so, he found that the young veterinary professor's method only put the tenacious anthrax bacilli to sleep. What Toussaint had failed to observe was that, after a time, the bacilli reactivated and regained their full lethal strength. Toussaint's vaccine was useless; in fact, it could be deadly.

Meanwhile Professor Colin, who had failed to prove his claim that *Bacillus anthracis* was not responsible for anthrax, now changed his tack and attacked Pasteur's anthrax experiments. Pasteur had recently conducted a study in which he'd concluded that chickens could not be infected with anthrax because of their high body temperature. Suspecting therefore that temperature had something to do with the virulence of anthrax, Pasteur had then tied test chickens to wooden boards to subdue them, and then cooled them in water before injecting them with the anthrax bacillus. The chickens, injected while their body temperatures were low, had died soon after being released.

But Professor Colin would not accept that they had been killed by anthrax, so Pasteur agreed to allow Colin to carry out an autopsy on the dead birds before a fascinated audience at the Academy of Sciences. Professor Colin's autopsy indeed confirmed that these chickens had died from the anthrax virus, as Pasteur claimed. But determined not to yield to Pasteur, Colin had promptly declared that these chickens had succumbed to the virus because they had been stressed by being tied to boards. Temperature and cooling, Colin said, had played no part.

Pasteur had a simple answer. With Colin and others watching closely, he injected three more chickens with *Bacillus anthracis*, tying them down but not cooling them. According to Colin's theory, the birds would be stressed by being tied down, and therefore succumb to the anthrax. Pasteur, convinced that the birds would not die because of their high body temperatures at the time of inoculation, gave the three test birds to Professor Colin to watch over. Both men knew that anthrax kills within forty-eight hours. For days and then weeks the professor kept reporting that he expected the test chickens to drop dead at any moment. They did die, but many weeks later, and not from anthrax – Colin humbly reported that they were eaten by a dog that gained entry to his chicken pen.

Pasteur, proven right yet again, returned to his anthrax research aided by Charles Chamberland and another new student assistant by the name of Emile Roux, working on the basis that Toussaint may have been onto something when he applied heat to the anthrax bacilli. Over the winter of 1880–81, Pasteur and his assistants experimented with a range of temperatures, and struck gold. Pasteur found that the bacillus' protective spores identified by Koch had an Achilles heel: in a narrow temperature band, the spores did not form. Just as winter was ending, Pasteur was able to announce that he had produced an anthrax vaccine by growing the microbe at a constant 42 to 43 degrees centigrade for eight days and infusing it with oxygen.

The secret of Pasteur's vaccine was that, in his own words, 'the anthrax microbe goes through different degrees of attenuation'.[14] The attenuated anthrax microbes produced via his method were neither too virulent to kill nor too weak to be useless as a vaccine. Most importantly, they would not regain their strength and become deadly again as they would if produced outside the 42–43 degree heat band. On 28 February 1881, Pasteur announced his breakthrough. His discovery would form the basis of a set of Pasteurian principles that would establish the foundation of modern immunology.

Pasteur's anthrax announcement caused great excitement in government, medical, scientific and agricultural circles. And great controversy, as was always the case when Louis Pasteur had something new to say. 'Instead of receiving the tribute of attention and admiration he deserves,' complained France's *Journal of Medicine and Surgery*, 'he encounters the frantic opposition of a few individuals of quarrelsome disposition, who are always ready to tear him down after listening as little as possible.'[15]

One persistent opponent of Pasteur and of vaccination, an elderly physician and Academy of Medicine member named Jules Guerin, had even challenged Pasteur to a duel the previous October after they'd had a war of words and Pasteur had shown that he was 'contemptuous of that ignorant criticism which doubts on principle'.[16] Pasteur had not accepted eighty-year-old Guerin's challenge with pistols, but he could not resist accepting another challenge from a different source in the spring of 1881, following his anthrax vaccine announcement.

Hippolyte Rossignol, a veterinary surgeon at Melun, was a Pasteur critic who did not believe that germs were responsible for disease. It was he who issued the latest challenge – he offered his own farm at Pouilly-le-Fort, six kilometres outside Melun, as the location for a public field trial of the new Pasteur anthrax vaccine, convinced that it would fail. Pasteur agreed without hesitation. Rossignol personally collected money to fund the trial, which would commence on 5 May, and the President of the Society of Agriculture of Melun, Baron de la Rochette, came to Pasteur on 28 April to agree trial conditions.

With the press and public taking great interest in what was shaping up as a decisive contest between Pasteur and his detractors, Pasteur's friends at the Academy of Science began to worry that he had painted himself into a corner, allowing no room for retreat if his vaccine were to fail for some reason that could not be detected in the laboratory. Pasteur's assistants Chamberland and Roux were equally worried. When Pasteur cabled them about the trial they immediately cut short their holidays and scurried back to the Rue d'Ulm laboratory to begin further experiments with the Pasteur anthrax vaccine. They were concerned that their master had been premature in accepting the Rossignol challenge, for the vaccine had not yet been fully tested. Pasteur had even agreed to Rossignol's late requirement that ten cattle also be included in the field trial, even though the vaccine had so far only been tested on sheep.

In Germany, Robert Koch had been following events ever since Pasteur had announced his anthrax vaccine discovery, and now he too expressed the belief that Pasteur was getting ahead of himself by going to public field trials.[17] Koch was in fact confident that Pasteur's vaccine would not work.

Young Adrien Loir, preparing to join his uncle fulltime the following year, would, when he saw Pasteur at Arbois in the summer, become privy to the goings-on in the Pasteur camp during the weeks leading up to the anthrax field trial and during the trial itself. It turned out that Chamberland and Roux were convinced that Pasteur's culturing method, using oxygen, had yet to be perfected and could not be relied upon. Earlier in April, prior to going on leave and before Pasteur agreed to the Rossignol trial, Chamberland had been testing a new version of the vaccine, attenuated with potassium-bichromate instead of oxygen. In a side-by-side test between the two methodologies on 13 April, Chamberland's potassium-bichromate vaccine had worked. But Pasteur's oxygen version had failed, killing its test sheep, with the anthrax regaining its full virulence once it was injected into the animal.[18]

In the first week of May, Pasteur, Chamberlain and Roux left Paris for Melun to carry out the field trial, taking sealed jars containing the anthrax vaccine and *Bacillus anthracis* with them. Chamberland and Roux were still worried that Pasteur had over-committed himself. 'When we remarked that the program was severe,' Roux later said, describing their departure from the laboratory that day, 'but that there was nothing to do except carry it out since he had agreed to it, Pasteur replied: "What succeeded with fourteen sheep in the laboratory will succeed with fifty at Melun."' In fact, said Roux, Pasteur was in unusually high spirits. Seeing the glum looks on the faces of his assistants, he grinned and jokingly urged them to make sure they didn't mix up the bottles containing the vaccine and the deadly anthrax bacillus.[19]

From Melun, the trio took a carriage to Rossignol's farm at Pouilly-le-Fort. A party of interested observers was waiting for them

– farmers, veterinarians, physicians and newspapermen, eager for the contest to begin. If Pasteur was right, all the test animals would both survive inoculation with the vaccine and survive subsequent injection with the deadly raw *Bacillus anthracis*. If he was wrong, either the vaccinations would kill them, or the bacillus would. As sheep, cattle and even two goats were produced for the trial, Pasteur appeared supremely confident. And there on 5 May, using a Pravaz hypodermic syringe – a metal instrument that took considerable time to load and use – Chamberland and Roux injected a mild dose of the vaccine into twenty-four sheep, six cows and a goat. Everyone then left.

Twelve days later the observers arrived to find all the inoculated animals alive and well. First points to Pasteur. With their audience forming a circle around them, Chamberland and Roux then injected the original thirty-one inoculated animals with a lethal dose of anthrax, and also injected twenty-five sheep, four cows and a goat that had not been inoculated on 5 May. At the insistence of Rossignol, who seemed suspicious that Pasteur might try to inject different material into the protected and the unprotected animals, the sheep and cattle were all given the anthrax injections alternatively – first an inoculated animal, then a non-inoculated one, and so on. Pasteur then invited the observers to return in forty-eight hours' time, and again the audience members withdrew, with Pasteur and his assistants adjourning to lodgings in Melun.

After twenty-four hours, Pasteur suddenly had a crisis of confidence. The tension of the past weeks had caught up with him. His assistant Roux saw his chief's demeanour change: 'Pasteur, who had rushed into the public experiment with such perfect confidence, began to regret his audacity,' Roux would write. 'For some moments his faith was shaken, as though he feared his experimental method might fail him.'[20] The weight of criticism from his opponents combined with the concerns of his supporters and those of his own assistants had finally fractured Pasteur's otherwise monumental self-belief.

Many of those critics had reminded Pasteur that he was, by training, nothing more than a chemist. 'I do not forget that medicine and veterinary practice are foreign to me,' he admitted during this period.[21] But by the next day, as the team set off from Melun for the Pouilly-le-Fort farm to view the final outcome of the trial, Pasteur had thrown off his doubts and was, Roux would observe, 'more assured than ever'.[22]

At the appointed hour on 2 June, a crowd more than 200-strong surged through Monsieur Rossignol's farm gate – this time, there were even members of the French Senate among the observers. As the crowd approached the two yards where the test animals were being kept, a great shout went up. The yard containing the non-inoculated animals was littered with dead sheep. Of the twenty-five sheep in that yard, twenty-one were dead, three would die very shortly, and the last would die that evening. The goat was also dead. None of the cows in the yard was dead, but all were clearly unwell. Displaying a large swelling where they had been injected, these cows had elevated temperatures and all had gone off their feed. In the yard next door, every one of the inoculated animals was a picture of health.

As Rossignol and other veterinary surgeons inspected the dead and living animals, scratching their heads, the majority in the crowd surrounded Pasteur and his assistants, congratulating them with cheers and handshakes and pats on the back. The demonstration had been emphatic. Pasteur was acclaimed 'the master' and Rossignol was never heard from again. Within weeks, Pasteur would be flooded with requests for his vaccine. The requests came from veterinarians throughout France who, in the words of Pasteur's son-in-law, had 'suddenly become fervent apostles of the new doctrine'.[23]

To meet the demand, Pasteur would establish a small manufacturing centre in an ENS annexe in the Rue Vauquelin, just around the corner from his Rue d'Ulm laboratory. By year's end, 33,946 sheep, oxen and horses would be vaccinated with the Pasteur vaccine. The following year, 1882, the number was 399,102.[24] A lucrative new income stream had been created for Pasteur and his laboratory.

What no one knew, and what was only revealed by Adrien Loir decades later and subsequently supported by Pasteur's own notebooks, was that at the last moment Pasteur had substituted his assistant Chamberland's potassium-bichromate version of the vaccine for his oxygen-attenuated version, and it was Chamberland's version that was used in the field trial. Back in February, Pasteur had announced that he'd created an oxygen-attenuated vaccine. Throughout March he'd repeated the claim, leading the trial observers and the world at large to believe for a generation or more that he'd used it at Pouilly-le-Fort.

The trial conditions that Pasteur had signed up for on 28 April hadn't specified the nature of the vaccine to be used, so as far as Pasteur was concerned there were neither practical nor ethical reasons to prevent him using the new version for the trial. Pasteur never lost his belief in the oxygen-attenuation method, and Loir would write that he returned to the oxygen method following the Pouilly-le-Fort trial and persisted with it until he perfected it. 'He was tenacious,' said Loir.[25]

Neither Pasteur nor Chamberland nor Roux ever revealed this 'Secret of Pouilly-le-Fort' as it came to be called.[26] The secret, and Pasteur's deception, did not alter the fact that Pasteur and his team had created an effective anthrax vaccine. And in the end the oxygen method prevailed. That Pasteur never mentioned the substitution at the time of the field trial indicates how determined he was to maintain his reputation. The field trial was seen as a spectacular success for him, a success reported around the world. He was not going to soil his triumph by admitting that the first version of the vaccine he'd boasted of had actually turned out to be a failure. Nor was he going to give ammunition to the man who was shaping up to be his greatest critic.

Robert Koch, the German doctor who'd first isolated the anthrax bacillus, and Pasteur were now locked in battle. In 1880 Koch, who was still only thirty-six, had been given a senior post with the *Reichs-Gesundheitsamt*, the Imperial Health Bureau, in Berlin where he would conduct ongoing bacteriological research with the aid of talented student assistants who would also make names for themselves in the scientific world. In 1881, both Louis Pasteur and Robert Koch had made presentations at the Seventh International Medical Congress in London where Pasteur spoke about his work on the attenuation of viruses. At the conference, the two men were introduced to each other by Joseph Lister. Koch was a Prussian, had even served as a doctor with the Prussian army during the 1870–71 German invasion of France, but Pasteur managed to set aside his hatred of the Germans to congratulate Koch on his work and to briefly shake his hand. It was the first and last time they would have such contact.

Shortly after both men had returned to their respective countries, the Imperial German Health Bureau published papers in which, in Pasteur's words, 'not only this discovery of attenuation but all my previous research on the microbes of disease was attacked with an odd brusqueness by Dr Koch and two of his pupils'. Pasteur did not answer these criticisms immediately. 'I wanted to reply when a favourable opportunity presented itself.'[27] That opportunity came following the June 1882 success at Pouilly-le-Fort and the subsequent publicity it generated. Pasteur was invited to speak at the International Hygiene Congress in Geneva in September. There he would answer Koch.

By this time, Koch's stature in the scientific world had grown enormously. For the previous March, while Pasteur was basking in the glory of his anthrax vaccine, Koch announced in Berlin that he had isolated and grown the bacillus responsible for tuberculosis, a disease without cure that killed hundreds of thousands of people every year. This was one of the greatest breakthroughs in the history of human medical research up to that time and made Koch a household name in Germany and well-known in medical circles worldwide.

When Pasteur took the stage in Geneva on 5 September, the meeting hall was packed with medical and scientific men eager to hear about his anthrax discovery. Sitting in the front row were Robert Koch and his pupils. As Pasteur delivered his address, thirty-eight-year-old Koch – balding, slight, with a walrus moustache and a neat, short beard – sat with folded arms watching the sixty-two-year-old Frenchman through wingless, gold-rimmed spectacles. Not only did Pasteur use the opportunity to detail his methods and successes, both of which Koch had dismissed, he invited Koch to debate him in front of the congress audience. Koch shook his head. In a brief response, the German haughtily announced that he would reply to Pasteur's address in writing. The congress ended and the delegates went home. Once back in Paris, Pasteur waited anxiously to see what his German rival would have to say.

In the late autumn of 1882, Adrien Loir arrived in Paris with his minor degree and joined his uncle's staff as his *preparateur* and secretary. Pasteur outfitted a loft bedroom for Loir in the ENS annexe in the Rue de Vauquelin, the facility being used for the manufacture of anthrax vaccine. Every morning, Loir walked around the corner to the Rue d'Ulm and went to work for his uncle in the laboratory. He would later say that Pasteur 'took complete possession of me. From the first day, I became his tool, the indispensable accessory that he would use as he saw fit'.[28]

Loir's first lessons from Pasteur were about the everyday procedures he must follow as his laboratory *preparateur*. First and foremost, he had to learn about protecting himself and the experiments from germs. Pasteur was fanatical about germs. At every opportunity he would lecture surgeons on the need to sterilise their hands and instruments before operating on patients. As incredible as it seems today, most doctors in the 1880s did not accept that germs caused infection, and very few used sterilising techniques or antiseptics. Eminent French surgeons opposed to Pasteur's germ theory would arrive in their operating rooms with unwashed hands, wearing the same soiled surgical gowns they had worn for previous operations. Some would deliberately conduct university lectures in their soiled gowns to send a message to Monsieur Pasteur that he should stick to chemistry while they took care of surgery.

Fire, said Pasteur, killed germs more certainly than anything else. 'In my laboratory, where I am surrounded by germs of every kind, I therefore never use an instrument without having first passed it through a flame.'[29] Then there was the thorough washing of hands with soap to counter germs. As Adrien Loir found, this went on throughout the day in Pasteur's laboratory, and according to a fixed ritual. What's more, 'Pasteur had a phobia about shaking hands, and that is probably why people found him haughty. He never held out his hand.' If he had no choice but to shake a visitor's hand, Pasteur would wash his hands as soon as possible after they left.[30]

To begin with, Pasteur treated Loir as his slave. 'It was only after he had constantly had me at his side for some time that it occurred to him that I should obtain some official credentials,' Loir wrote.[31] For a time Pasteur thought of sending the young man to pharmacy classes. Then he considered the veterinary school at Alfort on the outskirts of Paris, but studies there would have taken Loir away from the laboratory for too long. Pasteur himself had no medical training, and so by law could not carry out medical procedures on humans, not even as much as to administer an injection. It was probably for this reason that he decided that Loir should study medicine.

So, while young Loir worked with Pasteur in the laboratory every afternoon, in the mornings he went to work in the clinic of Dr Michel Peter, who shared a grandfather with Adrien Loir Senior – Joseph Loir, who had served as a veterinarian in Napoleon Bonaparte's army in Egypt. Pasteur and Peter argued regularly, for Peter could at times be among Pasteur's harshest critics. So it must have been at the insistence of young Loir's parents that their son trained under their relative Dr Peter.

Within weeks of arriving at the Rue d'Ulm, Loir found himself on a field trip to the south of France with Pasteur and his latest student-assistant from the ENS, twenty-six-year-old Louis Thuillier. Pasteur was now working on rabies, the vaccine for which would eventually bring him his greatest fame. He had a habit of dropping everything when something new came up and rushing off to distant parts equipped with his microscope and his curiosity. Such was the case in the middle of November 1882 when word reached the Rue d'Ulm from the Vaucluse region that there was an outbreak of swine erysipelas, the often fatal swine fever. While Pasteur was attending the conference in Geneva in September, back at the laboratory the brilliant young Thuillier had cultivated the swine fever microbe and quickly developed a vaccine. In November, Pasteur, Thuillier and Loir spent weeks in the Vaucluse district vaccinating pigs. When swine fever swept through the region a year later, not one of the vaccinated pigs fell ill.[32]

Pasteur arrived back in Paris from the Vaucluse expedition just before the Christmas of 1882 to find awaiting him a pamphlet issued by Robert Koch. Finally, after three months, Koch had deigned to reply to Pasteur's Geneva address. The more Pasteur read of Koch's pamphlet, the angrier he became, exclaiming that Koch, this upstart who had only 'come to science' six years before while he had been in the field for thirty-six years, was guilty of 'lofty disdain' and 'childish assertions'.[33]

What stung Pasteur most was that Koch accused him of having brought nothing new to science. According to Koch, Pasteur merely traded off the work of others. Koch would not even credit him with the creation of the anthrax vaccine. It wasn't only that Koch would not accept that vaccination was a legitimate tool of science and medicine. The German did acknowledge that a way of culturing an anthrax vaccine had been discovered, but by Toussaint, the young veterinary professor from Toulouse. According to Koch, Pasteur had merely improved on Toussaint's heat-treatment discovery and had disingenuously claimed the glory of the discovery for himself.[34]

On Christmas Day, while pacing up and down his Rue d'Ulm apartment like a caged lion, Pasteur, dictating to Adrien Loir, composed a lengthy and combative answer to Koch's response. 'What bias, Monsieur, and what errors in your account!' Pasteur raged. 'You attribute to me some errors which I had not committed. You refute them, then exult noisily.'[35] His answer was framed as an open letter. By the time Pasteur had finished, Loir had taken down what amounted to a small book. Point by detailed point, Pasteur aggressively countered Koch's claims and counter-claims. And it was obvious now why the German had waited three months to answer him. To attack vaccination, and Pasteur's use of vaccination, Koch had collected recent statistics on the stock vaccinated in France with Pasteur's anthrax vaccine. He then used these statistics 'in a most erroneous manner' according to Pasteur, skewing and misrepresenting some figures and omitting others to condemn vaccination. 'Entirely violent as your attacks are, Monsieur, they will not impede its success,' Pasteur declared. 'I must add, in conclusion, that everything predicts that preventative vaccination will be even more effective in the future.'[36]

Pasteur's open letter was published in the *Revue Scientifique* in Paris on 20 January 1883. The letter, and editorials commenting on it, were subsequently published in newspapers and journals around the world. Prior to this, Pasteur and Koch had been engaged in hit-and-run skirmishes. This letter was a declaration of all-out war between the two scientists. From that point on, the scientific, medical and veterinary communities became divided into two camps – those for vaccination, and therefore for Pasteur; and the anti-vaccinists who joined Koch in deriding and opposing vaccination. The divide became so marked that those who supported Pasteur used his word, microbiology, to describe his field of science while Koch and his supporters spoke of the very same field as bacteriology.

The war between Pasteur and Koch would be fought in many ways and in many places. But the most bitter campaign in that war would soon be fought in far-off Australia.

4

THE RABBIT INVASION

In the southern autumn of 1880, around the time Louis Pasteur announced to the world that he'd succeeded in cultivating a chicken-cholera vaccine, James Francis Cudmore was travelling up the Murray River on a paddle-steamer. The Murray, along with the Mississippi, Ganges, Yangtze and Nile, is one of the world's great rivers. Rising in the Snowy Mountains, it snakes 2530 kilometres, north, west and south until it empties into the Great Southern Bight near Adelaide. Along the way, it forms part of the border between New South Wales and Victoria. In the 1880s it was a major thoroughfare, plied by hundreds of shallow-draught river-steamers driven by stern and side paddlewheels.

Jim Cudmore lived at Glenelg, a well-to-do beachside suburb of Adelaide, and was riding one of these riverboats to visit a property of his near Paringa, 200 kilometres northeast of Adelaide. Jim was a 'squatter', so-named because in the early 1800s, squatters had literally squatted on unclaimed rural land that took their fancy; the government later regulated the system. By the 1880s, unlike 'selectors' – small-time farmers who struggled to make a living on 'selections' of no more than 128 hectares (320 acres) – colonial squatters like Jim Cudmore were wealthy men with vast land holdings. Jim controlled properties in South Australia, Victoria, New South Wales and Queensland that spanned hundreds of thousands of hectares. He owned only a small minority of that land; like all squatters, he leased the vast majority of his rural hectares from the government on short-term leases.

That autumn of 1880, as the river-steamer neared Paringa, Jim Cudmore noticed a small bundle of grey fur sitting on the western bank on the South Australian side of the river. At first Cudmore didn't believe his eyes. Looking more closely, he confirmed his first impression: it was a rabbit. The reason for his surprise was that he'd never seen a rabbit in this part of South Australia. He knew that rabbits were running wild in Victoria and causing havoc there, eating farmers' crops and stock feed, but the broad waters of the Murray had acted like a Great Wall of China, keeping the pests east of the river. 'I first thought the rabbit had been wilfully let loose,' Cudmore said later, recalling that day on the Murray. 'But since then I have had reason to change my mind. This rabbit evidently came from the scrub of Victoria, and was the forerunner of the wave of rabbits that followed.'[37]

Australia's first rabbits had arrived with the white settlers of the First Fleet in 1788. Imported during the early years of the colony of New South Wales and raised for its meat and fur and for felt manufacture, the European rabbit, *Oryctolagus cuniculus*, did not adapt well to the alien Antipodean environment. In 1821, farmer George Cox wrote from his property in the Hawkesbury district to his Sydney friend the Reverend Thomas Hassall about his caged domestic bunnies: 'The rabbits are not doing quite so well as we could wish, but for all that I shall not despair.'[38]

The credit, or discredit, for launching rabbits into the Australian wild and, in doing so, launching the continent's rabbit plague, is generally given to Victorian landowner Thomas Austin. The well-to-do English-born owner of Barwon Park, an 11,736-hectare property near Geelong in the colony of Victoria, Austin detested the local flora and fauna. This native of Somerset would form the Acclimatisation Society of Victoria with the objective of Europeanising the Australian landscape with imported plants and animals. As one contribution to that exercise, on Christmas Day 1859, Austin released twenty-four rabbits, five hares and seventy-two partridges on his property. He felt sure this could do little harm. In addition to providing a touch of 'home', it would offer 'a spot of hunting'.[39]

In ideal conditions, female rabbits ('does' as they are called) produce five or six litters each a breeding season, which in Australia runs through the colder months of April to October. Those litters range in size from two to eight rabbits, with a gestation period of just under one month between conception and birth. Young does can start producing from the age of four months. Austin's rabbits were released when the conditions happened to be perfect for rabbit breeding, with unseasonal summer rains making plenty of green grass available. In addition, on Barwon Park the released animals encountered few natural enemies among native animals or birds. Contrary to predictions by Austin's contemporaries, his rabbits flourished in the wild and multiplied.

'It was considered something of a feat at the time, nobody foreseeing the result,' commented Victoria's Chief Inspector of Stock, Edward Curr, nineteen years later.[40] Curr was referring to a wave of rabbits that, over the next decade, would expand progressively further north and west.

'The wave hit us in 1880, and kept on coming,' said Peter Waite, a director of the company that ran Paratoo Station in South Australia.[41] By the early 1880s, Victorian rural areas were overrun by millions of rabbits and the swelling bunny population had spread into New South Wales and South Australia. Tasmania, while an island, had not escaped the scourge; someone had released imported rabbits there too in recent times, and its farmland was also being overrun. And New Zealand had its own rabbit plague.

Not only were the invaders eating crops, they were eating the grass on which the graziers depended for their sheep, cattle and horses. 'Six or eight rabbits consumed as much grass as one sheep,' Victoria's *Weekly Times* reported, 'but that was not the extent of their robbery. They eat right down to the roots.'[42] Within a few years of spotting that first rabbit beside the Murray, South Australian squatter Jim Cudmore saw the holding capacity of his Paringa run and that of his next door neighbour drop from 147,000 head of sheep to 20,000 for lack of grass, all because of rabbits.[43] Other squatters reported their holding capacity reduced by half during this period, while on many marginal inland holdings, large tracts of land became incapable of supporting any stock. Farmers were using every method they could to combat the pest – hunting, trapping, poisoning and wire fencing. But they were losing the battle. Rabbit numbers continued to grow.[44]

In April 1881, a Captain Raymond arrived in Melbourne from New Zealand to demonstrate his horse-drawn Patent Grain Phosphoriser, the latest weapon in the farmer's anti-rabbit armoury. 'Some fifty or sixty gentlemen, including some of the leading stock-owners of the colony and New Zealand' watched as Raymond demonstrated how he mixed water, grain and phosporous in a boiler to create an effective bulk rabbit poison, then gave instructions on how and where to lay the poisoned grain for best effect.[45] Captain Raymond's phosphoriser was soon being employed throughout the colonies along with off-the-shelf rabbit poisons produced by colonial drug manufacturers: Presto Rabbit Poison, PPP Annihilator and Bosker Rabbit Poison would become household names in rural Australia.

Press advertisements for rabbit poisons at the time invariably showed bunnies shedding tears for their dead comrades, but it was the farmers of Australia who were shedding tears, for the bunnies were winning. Colonial governments were pressed by farmers' groups and parliamentarians from rural electorates to take the lead in the rabbit war. 'Either the rabbits or the men must go,' said Walter Madden, a member of Victoria's Legislative Assembly, in an impassioned parliamentary speech that was typical of the sentiments being expressed at the time. 'The land will not support both!'[46]

What finally jerked governments into gear was the threat from squatters to simply walk away from their runs once their leases ran out rather than waste another penny on fruitless rabbit extermination efforts. This was an era before personal income tax or company tax, and the sale and lease of Crown land was the single largest income source for the colonial governments of Australia. It was crucial to the colonies' fiscal futures that the exploitation of rural land continue.

By 1883, several Australasian colonial governments had legislated against the rabbit. In New Zealand, this kind of legislation had been introduced as early as 1877. In Victoria, a Vermin Board took up the assault against the local rabbit army. In New South Wales, via the new Rabbit Nuisance Act of 1883, the government of Premier Alexander Stuart compelled landowners to carry out rabbit extermination measures on their properties. A Rabbit Department was set up by the New South Wales Government to administer a system where the colony paid bounties for the scalps of all rabbits killed. Farmers were expected to pay for rabbit extermination on their own properties and then claim a rebate of up to three-quarters of the cost from the authorities. Government rabbit inspectors in uniforms and pith helmets rode around the farming communities of the southern colonies policing the rabbit cull and authorising bounty payments.

A whole new industry had sprung up overnight. Thousands of men flooded into the country districts to seek employment as 'rabbiters', as rabbit exterminators came to be known. At first, rabbiters hunted their prey with dogs and guns. Before long, saving bullets, they set rabbit traps. Then they set baits, and spread various poisons – phosphorised wheat, arsenic dissolved in potash and strychnine were common. An orchardist killed 500 rabbits in a single night with peaches laced with strychnine.[47] Warrens were filled in or carbon gas was pumped into them. Some hunters even rode down rabbits on horseback and netted them. Businesses opened in the cities to export rabbit pelts to the felt industry in Britain. One Sydney plant began to package tinned rabbit meat for the British market.

By 1884, the Victorian Government was lagging well behind the Governments of New South Wales and South Australia in its rabbit extermination spending. Its £6000 allocation – $2.4 million in today's money – paled in comparison to the £30,600 ($12.2 million) that was being spent by South Australia and the £74,000 ($29.6 million) by New South Wales that year.[48] This expenditure was seeing unprecedented numbers of rabbit scalps being handed in across New South Wales. Yet the plague still went unchecked.

With the winter of 1883 providing exceptionally good conditions for rabbit breeding, over the next few years one could drive from one town to another in western New South Wales across a veritable carpet of rabbits. At a property at Menindee, 100 kilometres east of Broken Hill, owner Herbert Hughes, a squatter who also had a sheep run in South Australia and a cattle station in Queensland, employed 108 rabbiters who killed 342,295 rabbits in just three months. Hughes had started paying his rabbiters 30 pence for each rabbit scalp. As the numbers caught swiftly became astronomical, Hughes had reduced the bounty to two pence a head. Elsewhere, the bounty would drop to a penny a scalp as the numbers of dead rabbits exploded. Despite the massive haul by Herbert Hughes' rabbiters, the pest continued to flourish on his property.[49]

With good summer rains in 1887, abundant green grass fed the ever-increasing rabbit horde. At the Momba Station at Wilcannia, owners Elder, Smith and Company spent up to £3000 ($1.2 million) a month on rabbit extermination that year, paying two-and-a-half pence per scalp. In just twelve months, 782,510 rabbits were killed on Momba and the neighbouring Mount Murchison Station.[50] Yet the colossal haul of scalps seemed not to dent the rabbit numbers around Wilcannia; bunnies continued to overrun both Momba and Mount Murchison stations.

At Langawirra Station, 120 kilometres east of Silverton in the far west of New South Wales where rabbits were spotted for the first time in 1884, the owners had rabbiters killing more than 40,000 a month by 1887. But the Langawirra cull made no difference to the plague. 'It was a failure entirely,' said Langawirra's manager, Alexander Bell. Not only were rabbit numbers on Langawirra undiminished, worse, the excellent stock feed promised by that year's solid rains had been destroyed. 'There is not a blade of grass to be seen,' Bell despaired.[51]

On Teraweynia Station, outside Condolbin, twelve government rabbit inspectors were supervising 161 rabbiters who destroyed half a million rabbits in three months during 1887. 'Yet no impression at all seemed to have been made,' said the New South Wales Rabbit Department's frustrated Superintendent, Harry Vindin. 'The rabbits were just as thick as ever.'[52] David Brown, manager and part owner of Kallara Station, a property covering a million acres (400,000 hectares) near Bourke in the northwest, which the rabbit wave had reached in 1883, had employed up to a dozen rabbiters a year between 1885 and 1887. Brown was of the opinion that absolutely no good came from the rabbit-hunting program.[53]

Another squatter, James Ormond, farmed 200,000 hectares near Wentworth in southwest New South Wales with several partners. Residing at a prime Collins Street, Melbourne address, not far from the Athenaeum Club, Ormond was a 'Collins Street farmer', as Victorians laconically nicknamed wealthy city businessmen who dabbled in rural pursuits. Ormond's dabbling at Wentworth had cost him £50,000 ($20 million) in improvements, but 1881, his first year in the wool business, had been extremely profitable, and had encouraged Ormond's hopes of continued success. By 1888, he was a disillusioned man, with wool prices only fifty percent of what they had been in 1881. 'Since then,' he complained, 'between the very bad seasons, the drought and the rabbits, these hopes have been shattered to pieces, and it has been a series of losses, one after another, year by year.'[54]

When finally in 1888 the colonial governments inquired into why so many rabbits could be exterminated and yet the overall numbers continued to grow and the plague continued to spread, the answer they received was 'human greed'. As Edward Curr, Victoria's Chief Inspector of Stock, was to remark, 'If you employ trappers then it becomes their trade and they live upon it.'[55] Put simply, without rabbits, rabbiters didn't have a job. It was in their interests for the plague to continue, and spread.

Prior to government susbsidisation of rabbit extermination, unemployed men would create work for themselves by covertly releasing rabbits in various areas and then going to the farmers of the district and offering to catch and kill the very rabbits they had set free on their land. Said Samuel Hubbe, South Australia's German-born Chief Vermin Inspector, 'I know that rabbits were liberated in 1874 in the Barrier Ranges in New South Wales, and Campbell's Creek on the Darling, close to Menindee.' He also believed this practice had been going on more recently in his own colony. 'Rabbits, to my knowledge, were liberated in Coringa and Kalloolloo, so that from these centres and others they spread out in an easterly direction towards New South Wales.'[56]

Once governments were paying for rabbit destruction, the rabbiters had a field day. The most common lurk was for the rabbit extermination parties to report an area free of rabbits and then strike camp and move on, making sure they left behind a few healthy bucks and does to breed. In this way, remarked Edward Curr, 'while the trappers are away the district gets as thick as ever'.[57] The rabbiters could come back again the following year once the rabbit population had again blossomed in a 'cleared' area, and start their work all over again.

Nor did the rabbiters care that, working in large parties as they did, they would scare rabbits from their burrows and herd the animals ahead of them. 'This had the effect of driving rabbits all over the country,' said Samuel Hubbe, 'to many places where they were never seen before – places which have since become thickly infested.' Rabbiters also created future work for themselves by keeping a few rabbits alive. 'In many cases I know of,' said Hubbe, 'they were wilfully transported – taken out of thickly breeding spots and planted in new ground. Of this I have absolute proof. I have also absolute proof that rabbits were scalped alive and then allowed to go in the hope that they would breed again.'[58] This way, the rabbiter was in a win-win situation – he was paid for the scalp, and the rabbit lived to breed again.

'The first thing the trappers kill is the natural enemies of the rabbit,' said John Reid, manager and part owner of Tintinallogy, a station that lay between Wilcannia and Menindee on the eastern side of the Darling River. Those natural enemies included the goanna, a large native lizard that whites then called an iguana; as well as native cats, feral cats and hawks.[59] The government at that time also paid a bounty for every hawk killed because hawks sometimes preyed on newborn sheep. This practice, said John Reid and others, encouraged the elimination of an enemy of the rabbit, taking away another aid to the destruction of the rabbit pest. Rabbiters especially disliked hawks because they would often eat the ears of dead rabbits before rabbiters could get to the carcasses and skin them. Rabbit ears were a vital part of the scalps for which the rabbit bounties were paid: no ears, no bounty.

Poisoning, Alexander Affleck believed, was much more effective than trapping. Rabbits are social animals and prefer to settle in mobs, making one specific locale their home. Affleck had found that when rabbits were settled and concentrated like this, instead of fleeing ahead of bands of rabbiters, they were much easier to destroy en masse with poison.[60]

As 1887 progressed, the farmers of New South Wales in their hundreds and thousands informed the government of the colony, some almost hysterically, that the current rabbit abatement program was not working. In its March–July 1887 sitting, parliament was gravely informed that the government's advisers believed that 'a large area of the Western pastoral country is so seriously affected' by the rabbit pest 'that the pastoral and agricultural interests are threatened with indefinite loss'.[61]

Few people in New South Wales, apart from rabbiters, believed that the compulsory trapping program had been anything but a costly failure. Vocal critics demanded that something be done to utterly exterminate the rabbit pest, something drastic and effective. And it had to be done at once, before the rabbit numbers grew from the millions into the billions, before the sheep stations of New South Wales became dust bowls, before the rural economy on which the colony depended collapsed.

5

SIR HENRY PARKES' RABBIT ERADICATION COMPETITION

A NEW GOVERNMENT CAME into power in New South Wales on 25 January 1887. And this change in government was to re-shape the war on rabbits.

Like the mother country, the colony of New South Wales had two houses of parliament. The Lower House, the Legislative Assembly, was elected. Members of the Upper House, the Legislative Council, were nominated by the leader of the government of the day, and sat for life. The Upper House had no real power, and was merely a house of review. The government was formed by the party with the majority in the Lower House, and that party elected a leader of the government. The official title of the leader of the government was Colonial Secretary, but by the 1880s the courtesy title of Premier of the colony, and sometimes even that of Prime Minister, was accorded the holder of that post.

In 1887, there were no formal political parties in New South Wales. In the style of the ancient senates of Greece and Rome, every member of the Legislative Assembly was, nominally at least, an independent. In practice, members fell roughly into two groups – those who favoured free trade with the other colonies and the world beyond, and those who advocated heavy import duties to protect local industries from foreign competition. The Free-traders, or Ministerialists, were in some ways like the Liberals, or Whigs, of Britain who also advocated free trade. The other side, the so-called Protectionists, were more like the Conservatives, or Tories, of British politics. At this time, the labour movement was in its infancy; the formation of the Labor Party of New South Wales – which would change the Australian political scene forever – was still four years away.

Because there were no formal groupings and no party machines, for years past it had not been unusual for some members of the Legislative Assembly to move from the Free-trade group to the Protectionist group, and back again, depending on the issues being voted on at the time. In theory this really was democracy as it was intended, with personal convictions playing a greater role than a party line. In reality, it made the New South Wales parliament a volatile mixture which, when shaken, could explode. With their majorities frequently evaporating, governments came and went with sometimes alarming regularity. This latest government taking office at the beginning of 1887 was the colony's twenty-fourth regime in thirty-one years. Numerous ministries over the years had lasted only a matter of months. One had notoriously come and gone in just two days.

With such a political background, it would seem a wonder that New South Wales had prospered: according to one estimation, the private wealth of the people of New South Wales was at this time £17 per head higher than that of Britain.[62] Yet for all its political comings and goings, this was no Latin American republic. Political debate in the bearpit that was Parliament House in Sydney's Macquarie Street could be vicious, the divisions bitter, and the feuds long-lasting; but the politicians of New South Wales respected the system that they swore to uphold, and respected the will of the voters. As one administration stepped aside and another slid into its place, the underlying bureaucracy remained unchanged, ensuring that the colony's stability was not threatened even as the pillars of political society occasionally tottered and fell.

The new Premier taking office at the beginning of 1887 was seventy-one-year-old Sir Henry Parkes, a gregarious man with a bushy white beard, long white hair and a knitted brow that gave him the look of a grumpy Father Christmas. Yet he was a formidable performer in the political arena, 'a shaggy lion', as Lady Jersey, the wife of one governor of New South Wales was to characterise him.[63] This was to be Parkes' fourth term as Premier of New South Wales. Described in his own time by the *Times* of London as 'the most commanding figure in Australian politics',[64] and by one of his fellow MPs as 'the last of those intellectual giants who laid the foundations of modern Australia',[65] Henry Parkes, after whom a New South Wales town and a Canberra suburb are named, was a colonial colossus, and one of the most colourful figures of all time in Australian politics.

A part-time poet himself, the English-born Parkes counted poets Lord Tennyson, Robert Browning and Australia's Henry Kendall among his friends, and regularly corresponded with British historian Thomas Carlyle and novelist Thomas Hughes, author of *Tom Brown's School Days.* A humanitarian with 'a genuine love of his country and his fellow man',[66] Parkes exchanged letters with Florence Nightingale for many years, and with her help he arranged for Nightingale-trained nurses to migrate to New South Wales to improve the local hospital system. The liberally minded Parkes also numbered Liberal Prime Minister of Britain, William Gladstone, his political idol, among his friends.

Parkes loved women, and women loved him. In an era when women were not allowed to vote, Parkes advocated female voting rights and encouraged women to attend his political meetings. He was to marry three times. When he came back into office in 1887 he was estranged from his first wife and was keeping a mistress, an attractive divorcee forty years his junior who had three young children.

The paradoxical Parkes was renowned for his management of government finances. When he departed office in 1883 he had left behind a government surplus of £2 million ($800 million). When he returned to power four years later, he found that the four free-spending governments that had reigned in the intervening period had turned his surplus into a deficit of £2.5 million ($1 billion) and blown out the government debt by £22,000,000, ($8.8 billion).[67] In 1887, he again had to set about turning a governmental profit to put the colony back on a sound financial footing.

Parkes came to New South Wales as a twenty-four-year-old; he and his first wife were migrants on an assisted passage. Educating himself by reading prolifically at night, Parkes went from labourer to newspaper editor in the 1840s. But it was politics that interested him most, and he was first elected to the parliament of New South Wales in 1854. In 1872 he became Premier of the colony of New South Wales for the first time, having previously served as a minister. Over the next nineteen years he would be Premier for a total of close to twelve years, making him by far the longest serving of the colony's leaders. In 1877, during his second term as Premier, he was knighted by Queen Victoria.

This was a time when neither parliamentarians nor government ministers were paid a penny; they were expected to provide their own sources of income while carrying out their parliamentary and government duties. Like British Prime Minster Benjamin Disraeli, another part-time author, Parkes fought a continuing and sometimes losing battle to keep his personal financial boat afloat, in contrast to his skills with the official purse. Yet despite these struggles, he remained strongly opposed to parliamentarians being paid. To him, it was an honour to serve the people. It wasn't that he lived extravagantly; Parkes' tastes were modest, his lifestyle frugal. He neither drank nor smoked, his clothes were unfashionable and ill-fitting. But he was a soft touch for those seeking money. 'I have never withheld my last shilling from those who needed it more than I,' he once said.[68]

By early 1887, within months of returning to office, Premier Parkes was again without income and technically insolvent after the last of his dozen heavily mortgaged rental properties had to be auctioned off at giveaway prices. Because he had just come into government, and with a record parliamentary majority at that, Parkes' friends were able to make an arrangement with his creditors which meant that, in return for assigning his entire estate to his creditors and walking away with just the clothes on his back, Parkes was not again declared bankrupt and could continue to sit in parliament and run the colony. Among the assets he lost was the summer retreat he loved – 'Faulconbridge', a cottage on a plot hacked from the bush in the lower Blue Mountains where he kept a menagerie of animals.

In November 1887, the friends of Sir Henry Parkes would call a public meeting at Sydney Town Hall to launch a benevolent fund designed to relieve him of future financial worries. A huge crowd of Parkes' supporters would turn up, and a total of £9,000 ($3,600,000) in donations would soon be collected in Sydney and around New South Wales. The Parkes' benevolent fund was expected to generate annual interest of £540 ($216,000), on which, it was hoped, Parkes and his family would be able to live in prime ministerial comfort while he devoted himself to furthering the colony's prosperity.[69]

His personal financial crisis was at its zenith when he'd thrown himself into the election campaign of 1886–87. Despite his age, Parkes was in fine political fettle. 'He was a born politician,' said fellow Free-trade MP and Parkes' supporter, William Walker.[70] Parkes made this the most vibrant political campaign in the colony's history. Travelling with apparently boundless energy throughout the colony he spoke at one political meeting after another, giving his support to other candidates who shared his views, many of them young men whom Parkes felt showed political promise. The 'robust' and 'imposing' seventy-one-year-old drew vast crowds wherever he spoke.[71] 'He loved popular applause,' said Will Walker,[72] and during the campaign Parkes generated deafening applause and rollicking laughter as he entertained his audiences with cheeky, quirky, and sometimes controversial allusions to his opponents. In one 1887 campaign speech to a packed hall at Liverpool, southwest of Sydney, he described those opponents as 'withered tarantulas', 'miserable poodle-headed creatures', and 'blacklegs, fools and anarchists'.[73]

Political adversaries such as C.O. Waldow complained: 'His wonderful power of oratory, his ready wit and his venomous attack upon his opponents sway the masses to and fro.' But, he scornfully added, 'He possesses no ability beyond the power of representing the most glaring falsehoods as the purest of truths.'[74]

As it happened, the masses wanted to be swayed. The voters of

New South Wales were sick and tired of the comings and goings of one government after another; four in four years were three too many for them. City people were worried about urban unemployment, country people were concerned by the downturn in the rural economy. And people everywhere jumped onto the Free-trade bandwagon, with its proponents' promise of prosperity. But most of all, they came out to vote for Sir Henry Parkes. 'It was remarkable how the constituencies turned towards him, and hailed him as the statesman and the leader that was wanted,' senior civil servant and Parkes admirer Charles Lyne observed.[75]

The unprecedented public support for Parkes had a rub-off effect. 'Very quickly,' said Lyne, 'it was recognised by candidates for the new parliament that supporting Parkes was the high road to success, and disapproving of him, certain defeat.'[76] Candidates everywhere urged voters to support them for no reason other than that they supported Sir Henry Parkes. Parkes won his solid majority. The wave of popularity that swept Parkes into power swept away scores of Protectionist supporters of defeated Premier Sir Patrick Jennings, and when the new parliament sat for the first time on 25 January 1887, new Premier Sir Henry Parkes and the Free-traders could count on the votes of eighty-three members of the Legislative Assembly as opposed to the forty-one members now sitting on the benches of Her Majesty's Loyal Opposition.

Parkes saw this crushing victory and huge majority as a mandate for change and launched into a challenging legislative agenda. The period of this fourth Parkes ministry, said Charles Lyne, 'constituted one of the most important and interesting periods in the colony's history'.[77] Parkes and the nine members of his cabinet had an 1887 budget of £38 million ($15.2 billion) to administer, with New South Wales' army and navy and their close to 6000 mostly part-time personnel, together with the colony's police force, jointly accounting for the largest slices of the colonial cake.

The Premier's first instruction to his cabinet was for economies to be made. The Treasurer was ordered to deliver a surplus, and a royal commission was quickly set up to inquire into the civil service and recommend efficiencies. Despite the attention to economies,

Parkes pressed forward with new initiatives in education and the way the colony's railways were run. He promised government-sponsored work to help the unemployed. And he passed an amended Land Act that secured tenure for pastoral tenants, which both pleased the many squatters whose leases were due to expire in 1888 and increased Government income from Crown land rents.

Once his other agenda items had been addressed, the Premier turned to the matter of rabbits. The minister responsible for the New South Wales Rabbit Department was the Secretary for Mines – not because rabbits make burrows, which might be considered small mines, but because this minister was also responsible for the Agriculture Department. The Secretary for Mines in the new Parkes ministry in 1887 was thirty-seven-year-old Francis Abigail. Born in London, the son of a grocer, Frank Abigail had worked in a boot factory before opening his own leather business in George Street, Sydney. After Sir Henry Parkes had seen his potential and encouraged him to go into politics, Abigail had been elected to the Legislative Assembly as a Free-trader in 1880. His seat in Parkes' 1887 cabinet was his first ministerial appointment. A teetotal Congregationalist, active, energetic and passionate, Abigail was an enthusiast who, once he took up a cause, threw all his energies into it.

As the first busy session of the new parliament was drawing to a close in July 1887, Sir Henry Parkes and his Mines Secretary sat down to discuss the rabbit problem. In Sir Henry's spacious office in the Colonial Secretary's Department, on the corner of Macquarie and Bridge streets in the administrative heart of Sydney, they discussed various options. Parkes was looking to save money in the Rabbit Department budget and he and Abigail were in absolute agreement that the scheme introduced by the Government of Premier Alexander Stuart providing for the payment of bounties for rabbit scalps had proven an enormously costly failure.

It was decided that in the next parliamentary session Abigail would amend the provisions for the bounty scheme to terminate it after 1887. This would save the Government tens of thousands of pounds, but it would not solve the rabbit problem. By the end of 1887, more than twenty million rabbits would be killed in New

South Wales in that year alone.[78] Yet the culling barely seemed to dent overall rabbit numbers. With perhaps a billion rabbits running wild in Australia and New Zealand, a more wide-reaching and permanent solution was needed.

As the pair discussed other potential solutions to the rabbit plague, Abigail remembered a speech given the previous year by Dr John Creed, a Woollahra general practitioner and editor of the *Australasian Medical Gazette* who was also a free-trade-supporting member of the Legislative Council. In that speech, Creed had recommended that the previous government consider offering a prize of £50,000 ($20 million) for the person or persons who could provide a contagious biological remedy that would wipe out the rabbits plaguing the colony without affecting other domestic or native animals. Such an apparently simple remedy had seemed far-fetched to the unimaginative Jennings ministry, and nothing had come of Creed's suggestion. Abigail proposed that perhaps they should give serious consideration to the Creed idea, and Parkes, after thinking about it for a moment, agreed.[79]

Neither of these two politicians had scientific or medical training but Parkes was well read and Abigail, who was on the board of Sydney Hospital, would have overheard doctors discussing the recent developments in biology in Europe. Between them, the two men had enough exposure to the subject to agree that the Creed proposal had distinct possibilities. But they considered a £50,000 prize too much. Surely, they thought, a prize of £25,000 ($10 million) would be more than enough to attract the attention of the world's leading figures in the area of biological research? Compared to the New South Wales Government's expenditure under the Rabbit Nuisance Act between 1883 and 1887 of £577,088, ($230 million) with 1887 accounting for close to forty percent of that expenditure,[80] a prize payout of £25,000 was a bargain, especially as a successful outcome to the competition should mean no further money need be expended on the eradication of rabbits.

Such a lucrative competition could be expected to attract a quota of crackpots and charlatans, so the Premier and his Mines Minister agreed that the prize would only be paid once the winning entry had been successfully tested for a full twelve months. Parkes instructed Abigail to prepare an advertisement for placement in all the leading newspapers of the world, particularly in Britain, France, Germany and the United States, calling for entries in the rabbit-remedy competition. At worst, Parkes and Abigail would have told themselves, if the competition produced no worthwhile results then all the government would lose would be the cost of a few newspaper advertisements. On the other hand, if the competition yielded the desired result, Parkes and Abigail could expect to be hailed as heroes in New South Wales for eradicating the rabbit pest.

With all the other matters on his mind at the time, it's likely that Parkes forgot all about rabbits the moment Frank Abigail hurried enthusiastically from his office to launch the project. Neither man could have anticipated the international controversy they were about to create with their rabbit eradication competition.

6

PAYING FOR PASTEUR'S INSTITUTE

FOR YEARS, LOUIS PASTEUR had been attempting to gain French Government support for an institute devoted to microbiological research and vaccine manufacture. In December 1882, after Pasteur, Charles Chamberland and Emile Roux had developed the anthrax vaccine, Pasteur had proposed to the Council of Ministers that the government set up an institute to take over the manufacture and sale of the vaccine, with Pasteur employed as its director and Chamberland and Roux as deputy directors. But the government had turned down his proposal.[81] Hence, Pasteur had personally undertaken the manufacture of the anthrax vaccine, selling it throughout France via a Paris-based commercial agent. The profits from these vaccine sales went into the pockets of Pasteur, Chamberland and Roux. Much of Pasteur's share was ploughed back into his work, but he was also able to put aside a tidy sum. Three years after developing the anthrax vaccine, when he made his breakthrough with rabies, once again the idea of establishing a public institute headed by Pasteur was raised.

Pasteur had begun studying rabies in late 1880 after a physician at the Troussou Hospital had sent him mucus samples from the palate of a child who'd just died from the disease. Humans contract rabies after being bitten by a rabid animal and usually die from it. Pasteur inoculated a pair of rabbits with the mucus and they died within thirty-six hours. In the blood of the dead rabbits Pasteur found the bacillus responsible for the disease. But all efforts to attenuate an effective rabies vaccine over the next few years had come to nothing. The more the solution eluded Pasteur, the more he wanted to know about the disease, conducting all manner of tests. 'Even now the experiments are in full swing,' said his son-in-law René Valéry-Radot in 1883. 'Biting dogs and bitten dogs fill the laboratory.'[82]

Pasteur's obsession was driven by his rivalry with Germany's Robert Koch. Pasteur had seen Koch receive universal acclaim for his work with human diseases – Koch's 1882 discovery of the cause of tuberculosis had made his name known to millions of people around the world. A year later, Koch had beaten Pasteur in a race to discover the cause of cholera. As Koch's German team succeeded in isolating *Vibrio cholerae*, Pasteur not only lost the race but also one of his favourite young researchers, Louis Thuillier, to the disease. So Pasteur, passionate about defeating rabies, continued his experiments for another two years, with many of the tests carried out by young Adrien Loir and a new Pasteur student assistant, Eugene Viala.

The rabies breakthrough came when Pasteur's chief assistant Emile Roux became involved, and in late 1885 Pasteur announced to the Academy of Sciences that he and his collaborators had succeeded in attenuating the rabies virus in the form of dried spinal cord from rabies-infected rabbits. When this dried material was mixed with bouillon and injected into healthy test animals, it created immunity from subsequent disease when the active disease was injected into a vaccinated subject. In addition, because rabies has a long incubation period, taking between thirty and ninety days to cause death depending on how far the bite or injection site is from the brain, Pasteur realised he could vaccinate a victim shortly after they had been infected, and save them. For the first time, Pasteur was working not only with a preventative, but with a cure.

Following his first human trials, Pasteur was besieged by rabies victims seeking the new Pasteur cure. In their hundreds and then their thousands they came, not only from France, but from as far away as Russia and the USA. As more and more rabies sufferers arrived in Paris, Pasteur was joined by Dr Jacques-Joseph Grancher, head of the paediatric unit at Paris's Children's Hospital, to take charge of free morning vaccination sessions for all comers, assisted by young Loir and Viala. During this period, Adrien Loir even injected himself with rabies and the Pasteur vaccine. Dr Grancher had accidentally injected himself with rabid material and had to take the cure; in solidarity with Grancher, Pasteur had wanted to also receive the vaccination. To prevent the master from taking any risk with his health, Loir and Viala had volunteered to receive the vaccine in Pasteur's place. When Loir's sensitive mother later learned about this, she fainted.[83] But Loir, Viala and Grancher all came through without any ill effects.

Demand for the free Pasteur rabies cure exploded. With 2500 patients in the first twelve months, Pasteur had to recruit two medical students to help Grancher with the rabies program, and a surgeon to attend to patients' bite wounds. Pasteur paid the cost of this rabies inoculation program from his own pocket, in part using the profits from anthrax vaccine sales. But this could not continue. It was costing Pasteur a small fortune, and the hundreds of people thronging to his laboratory each morning – 'every one of them with the vague terror hanging over him or her',[84] in the words of American physician and author Oliver Wendell Holmes who visited Pasteur in the summer of 1886 – disrupted Pasteur's other research.

Money had always been a concern for Pasteur. In 1874, France's National Assembly had granted him a state pension for life of 12,000 francs. This is the equivalent of just $19,200 in today's money. Recognising that common labourers earned more than this, in 1882, after his anthrax discovery, the National Assembly increased Pasteur's annual pension to 25,000 francs ($40,000), still not a fortune by any means. The ENS provided for Pasteur's laboratory and living quarters rent-free but he had to meet his own expenses for food, clothing, domestic help and Pasteur's considerable travelling expenses, as well as the expenses of his family and of secretary/assistant Adrien Loir. The anthrax vaccine profits generated since 1882 helped, but despite his fame Pasteur was not a well-to-do man.

By early 1886 it was clear that the tide of rabies patients would only increase and that a permanent clinic would have to be created to handle them. Meanwhile Robert Koch, by now Professor of Hygiene at Berlin University, had only months before presided over the opening of the government-funded German Institute of Hygiene in Berlin as its first director. The new German institute would investigate a range of human diseases and train bacteriologists: once again the Germans had stolen the march on the French.

Pointing out that the German Government was funding microbiological work by German scientists, Pasteur now petitioned the French Government for help with his burgeoning rabies clinic. But as had been the case when Pasteur had proposed less than four years earlier that a government institute be built to manufacture anthrax vaccine, no state money was forthcoming. A temporary home was offered in an empty government building on the far side of the River Seine. This would relieve immediate pressure on accommodation, but it was a far from satisfactory long-term solution. The move to the new premises was, for Pasteur's assistants, a disruption, but a necessary one. 'We had to leave the hospitable Rue d'Ulm to establish ourselves on larger grounds borrowed from the former College Rollin,' said Emile Duclaux. 'It was while we camped there that the international subscription was opened.'[85]

That subscription was initiated by Pasteur during a session of the Academy of Sciences on 1 March. After delivering a paper containing the latest information on his rabies treatment program, Pasteur declared, 'This calls for the creation of an antirabies vaccination establishment.'[86] The proposed institute would have him as its director and would also employ Roux, Duclaux and Grancher as his deputies – his former assistant Chamberland had been elected to the National Assembly as a deputy for the Jura *département* the previous year. 'In the Paris establishment we would of course have to train young scientists, who would then take the method to faraway countries,' Pasteur told the Academy.[87]

The Academy session enthusiastically endorsed the idea of an antirabies institute and a committee was set up to promote a public subscription that would raise the many millions of francs required to turn Pasteur's idea into a reality. The committee was chaired by Admiral Jurien de la Gravière, the commander of France's Mediterranean Fleet during the Franco-Prussian War who was now an author of military books and a member of the French Academy. Pasteur's colleague Grancher became secretary of the committee.

Back in January, Pasteur had received a cheque for 40,000 francs from a wealthy philanthropist, Count Laubespin, and Pasteur contributed this and 100,000 francs from his own savings to launch the subscription fund. Eleven of Paris's twenty newspapers and countless regional papers quickly lent their support to the campaign and solicited contributions from their readers. The government still did not offer any money but both it and the Banque de France offered to help with the collection of public subscription collections. Donations could be sent to the bank, to public treasurers general, to tax collectors, and to private collection agents who would pass the money onto the Credit Foncier, which held the growing fund.

The name of Pasteur's new institute was the subject of considerable discussion. Numerous possible titles were debated. Pasteur wanted it to be the Rabies Institute of France or something similar, but friends and colleagues felt that such an institute should do more than just carry out rabies treatments; it should, they believed, be involved in broad microbiological research and teaching. Pasteur did not disagree, nor did he disagree when Duclaux and others proposed that it be called the *Institut Pasteur*. It was reasoned that such a title, incorporating the name of the most famous microbiologist of them all, would help with the fundraising, which it most certainly did.

Through March and April an avalanche of francs came tumbling in from around the country. It was not only the affluent who contributed – one newspaper reported that a gendarme gave one franc, a poacher, half a franc. A total of 48,365 francs was collected in German-occupied Alsace-Lorraine where the local press reminded their readers that it had been a local boy, Joseph Meister, who had benefited from the very first Pasteur rabies treatment. From Germany itself, just 105 francs came in. Donations were received from other parts of Europe, from America and beyond. Emperor Pedro II of Brazil sent 1000 francs. A newspaper in Milan, Italy raised 6000 francs. In May, the Scentia Society organised a charity gala concert at Paris's Trocadéro Palace to raise money for the fund. Pasteur attended and gave a speech of thanks, saying that he had not attended a theatre like this more than ten times in his life.[88]

But within a matter of months, the initial public enthusiasm waned as the newspapers found new novelties to write about and donations dried up. Pasteur wrote to a friend on 24 June that the subscription was 'going well' but had a long way to go before it met its target. 'The Credit Foncier, which acts as a central collection agency for the subscription, will have taken in between 1,400,000 and 1,500,000 francs,' Pasteur said. 'But we will need two or three times as much. A considerable capital must be invested if there is to be a sufficient annual income to pay for materials and personnel.'[89] Pasteur had even resorted to the exhausting and, for him, embarrassing fundraising method of personally visiting potential wealthy sponsors and asking for money. Sometimes he was successful, other times not.

In October, Adrien Loir came back from a visit to Russia with another 97,839 francs from Czar Alexander III of Russia himself. Six months earlier, Prince Alexander of Oldenburg had appealed to Pasteur to help the Russians set up an antirabies laboratory in St Petersburg. By this stage Pasteur trusted Loir implicitly; his skills, loyalty and discretion made him the perfect Pasteur ambassador. On 14 July, Bastille Day, twenty-two-year-old Loir had set off for Russia with a cage containing two rabid rabbits, accompanied by ENS student Charles Perdix who would serve as his assistant. Loir duly set up the laboratory in St Petersburg and handed it over to the Russians, returning to Paris with the prince's thanks and the czar's cheque. It would not be the last time that Loir served as Pasteur's foreign troubleshooter – the following year he would embark on missions to Italy and Denmark.

Despite Loir's successful Russian expedition, by the time 1887 came around the subscription fund was still well short of Pasteur's target of between 3 million and 4.5 million francs. Even though the Pasteur Institute did not yet exist, Pasteur authorised Duclaux to produce a monthly scientific journal called the *Annales de l'Institut*

Pasteur, and the first issue was published in January 1887. To this day the Pasteur Institute dates its beginning from January 1887. Three months after the inaugural publication of the *Annales*, a second publication, the *Bulletin medical*, was launched by the Pasteur team. Edited by Grancher, this journal would expend much ink responding to attacks on Pasteur and on vaccination which continued to appear in the French and foreign press.

As always, Pasteur was impatient to move forward and in March 1887 a site for the new institute was purchased with money from the subscription fund. Pasteur was not in love with the location at first. He had wanted his institute to be in the heart of Paris, preferably on the Left Bank where he had trained, lived and worked for much of his life, close to the ENS. The pragmatic Emile Duclaux was able to convince Pasteur that the available Left Bank site that Pasteur had his eye on was both too small to allow for future institute growth and much too expensive for their means. After resisting the idea for some time, Pasteur agreed to a site at Grenelle, some distance southwest of the city centre.

At that time Grenelle was little more than a village on the urban fringe but Duclaux could foresee the day when the city would spread around it. The Grenelle site, on the narrow Rue Dutot, was flat and spacious although marshy. Best of all, it was relatively cheap – 420,000 francs.[90] In 52 BC, here on the plain of Grenelle, Julius Caesar's efficient deputy, Titus Labienus, had led four Roman legions in a bloody victory over tens of thousands of Gallic warriors. Now Louis Pasteur, the son of a Gallic soldier, planned to build an institute designed to save lives on the site of that two thousand-year-old battle.

Sixty-four-year-old Pasteur had been worn out by his own battles over the past year – a year of intense rabies research and sometimes controversial treatments, a year of regular attacks on his science at home and abroad. All this, plus the struggle to raise funds for his institute. In December, he suspended his research and left Paris for six months. Over the winter of 1886–87 he, his wife, daughter, son-in-law and grandchildren stayed at the sumptuous villa of a rich admirer at Nice, on the Riviera. Grancher was left in charge of rabies treatments at the College Rollin, Loir ran the laboratory at the Rue d'Ulm and Duclaux continued producing anthrax vaccine at the Rue Vauquelin facility.

During this long and welcome break, Pasteur planned the buildings that would serve as his monument. In his youth he had proved to be an accomplished artist, as his portraits of his parents and other Arbois residents testify to this day. And now he sketched his ideas for an institute complex to guide architect Félicien Brébant, who had offered to draw the plans for the institute free of charge. Pasteur envisaged an institute built in stone in the grand, chunky Louis XIII style. There would be two identical main buildings, each of four storeys and quadrangular in shape, the two linked by corridors at all four levels. At the rear would be stables and servants' quarters. A broad flight of stone steps would lead to the front door of the main building. A large, airy entrance hall would open onto administrative offices to the right while to the left there would be a grand, wood-panelled library with a massive oak ceiling. The building would contain rooms for rabies treatment, waiting rooms and recovery rooms all looking out onto restful gardens. There would be five separate laboratories, each working in a different field, plus photographic darkrooms and classrooms. The complex would be staffed by five department heads and fourteen assistants.

The top floors would be reserved for private quarters. Here would be spacious apartments with high ceilings and large windows overlooking the gardens. These apartments would be for Pasteur and his wife, and where Roux, Duclaux, Grancher and Loir would also come and live so as to be within a minute or two's walk of their workplaces. Compared to the facilities to which Pasteur had been accustomed all these years, Pasteur's new institute would be a scientific palace.

In April, Pasteur returned to Paris. An earthquake in February had disrupted the idyllic peace of the Riviera and Pasteur and his family had left Nice and gone to Arbois for another three quiet months. During his absence from the capital, the definitive statute of the Pasteur Institute, its charter, had been drawn up, and now that he was back in Paris he had it officially registered. The statute established the institute's role: the treatment of rabies by the Pasteur method, and the study of virulent and contagious diseases.

The statute also provided for a director and a twelve-member board plus an assembly of thirty members who would appoint a new board every three years. Louis Pasteur was to be the institute's first director, appointed for life. His successors would be appointed for six-year terms. To show the national significance being accorded the institute, the statute was signed by leading French journalists, by the secretaries of all five academies that made up the Institute of France, as well as by representatives of the Faculty of Medicine at the Sorbonne, of the French Senate and of the Chamber of Deputies. Notably, it was also signed by the heads of the banks Credit Foncier and the Credit de France.

Pasteur was still not greatly enamoured with the Grenelle site which he considered isolated and less than prestigious, and when, in June, the institute's small foundation stone was laid, the act was performed by Emile Duclaux. But that was as far as construction work went. On 4 June, the President of the French Republic, Jules Grévy, decreed that the Pasteur Institute was to be considered an institution of 'public usefulness'. The Government apparently wanted to have its cake and eat it too – it would still not put any money toward the creation of the institute, but now the Ministry of Commerce and Industry would have an official oversight role in its development, and in its work once it was up and running.

Clearly, the government was concerned that the fundraising for the Pasteur Institute was dragging its feet. It would be a national embarrassment should the Pasteur Institute fail to go ahead after all the publicity and patriotic fervour surrounding the establishment of the subscription fund, and more recently over the signing of the institute's statute. Foreign donors such as the Emperor of Brazil were asking about the progress of construction. Czar Alexander of Russia had expressed interest in sending high-ranking representatives to attend the institute's official opening. In the worst-case scenario, Jules Grévy's government would have to step in and take over the project. But not through any desire to help Pasteur personally.

President Grévy was no friend of Pasteur's. An avid republican, back in 1877 he had opposed imperialist Pasteur when the scientist decided he would run for the Senate. Pasteur had not overcome the disappointment of having the Senate seat conferred on him by Napoleon III in 1870 snatched from his grasp by the Franco-Prussian War and the subsequent change of government. So, seven years later, thinking that his renown as a scientist would serve him well, Pasteur had run for the Senate as a representative for the Jura. In opposing Pasteur in 1877 Grévy, a fellow native of the Jura and at that time President of the National Assembly's House of Deputies, had cleverly used Pasteur's fame against him, telling the voters that Pasteur was much too valuable to the nation to waste his time and talents gathering dust in the Senate. When the votes were counted, Pasteur came last of all the candidates. Ever since, Pasteur had been barely able to hide his disdain for Grévy.

Over Pasteur's head now hung the threat that Grévy's government would step in and take control of the Pasteur Institute project, with Grevy's 4 June announcement as the first step down that road, giving the government the necessary excuse to act should worse come to worst. If the government did step in, Grévy would make sure that Pasteur was sidelined to the role of a powerless figurehead, and Pasteur knew it.

The architect's plans, which Brébant drew to Pasteur's design, were soon approved. But still there was no meaningful movement at Grenelle. Through the summer and into the autumn of 1887, pressure mounted on Pasteur. From the government and from both friends and enemies in high places came questions about the state of the subscription fund. It was still at least a million francs short of where it should be. The bankers were prepared to advance sufficient funds to allow the construction to proceed only if they knew that the project would be completed, and only if Pasteur could provide sufficient collateral to cover the shortfall. Pasteur had no such collateral. He had no assets, and had exhausted his savings. From where then, would the money come?

Pasteur had exploited every conceivable avenue for adding to the subscription fund. The question of the financial viability of the

Institut Pasteur, Pasteur's son-in-law was to say, was weighing heavily on Pasteur's mind.[91] As the pressure continued to grow, the ogre of Jules Grévy hovered in the wings. If Pasteur could not soon find a solution, his dream would be shattered; he would lose control of his institute.

The strain told. On 23 October, while visiting his daughter Marie-Louise, Louis Pasteur suffered another stroke.

7

THE RABBIT ERADICATION PRIZE

EVEN BEFORE HIS LAST stroke, to save Louis' eyes from strain Marie Pasteur had formed the habit of reading newspapers and journals to her husband some nights after dinner. Emile Roux later said that Marie was not only an incomparable mate for Pasteur, she was also his best collaborator.[92] Among the many ways she helped her husband, Marie, like a good press agent, would scan a multitude of publications looking for mentions of Pasteur or items that might be of interest to him, for reading to him at night. One paper that generally received her attention was *Le Temps*, a Paris daily. Other papers, like the popularist *Le Siècle*, had a larger circulation but, with novelist Anatole France as *Le Temps'* literary critic and with contributions from leading writers and scientific figures of the day, the better-educated French readers and the foreign diplomatic corps alike considered *Le Temps* a quality paper.

On the evening of Saturday, 26 November 1887, the Pasteurs' son, the now thirty-six-year-old Jean-Baptiste, and his wife of thirteen years, Jeanne, had joined the Pasteurs at their apartment at 45 Rue d'Ulm for dinner. Serving at the French embassy in Rome during this period, Jean-Baptiste was home on leave. After the meal, they sat in front of the fire while Marie read from *Le Temps* as usual, with

Jean-Baptiste, Jeanne and the weak, grey-faced Louis who was slowly recovering from his second stroke in nineteen years only the month before, listening intently and occasionally making comments.

The latest front-page news was satisfying to Pasteur, with the son-in-law of President Jules Grévy caught trafficking in the decorations of France's revered *Legion d'honneur*. Within days, the scandal would force Pasteur's old foe Grévy to resign the presidency. But it was an advertisement from the 9 November issue of *Le Temps* that really grabbed Pasteur's attention:

PROCLAMATION BY THE GOVERNMENT OF NEW SOUTH WALES
EXTERMINATION OF RABBITS
Department of Mines, Sydney, 31st August, 1887.

It is hereby notified that the Government of New South Wales will pay the sum of £25,000 to any person or persons who will make known and demonstrate at his or their expense any method or process not previously known in the Colony for the effectual extermination of rabbits, subject to the following conditions, viz:

1. That such method or process shall, after experiment for a period of twelve months, receive the approval of a Board appointed for that purpose by the Governor with the advice of the Executive Council.
2. That such method or process shall, in the opinion of the said Board, not be injurious, and shall not involve the use of any matter, animal or thing, which may be noxious to horses, cattle, sheep, camels, goats, swine or dogs.
3. The Board shall be bound not to disclose the particulars of any method or process, unless such Board shall decide to give such method or process a trial.

All communications relative to the preceding should be addressed to the Honourable F. Abigail, Secretary for Mines.

FRANCIS ABIGAIL
Sydney (New South Wales)[93]

Louis Pasteur was suddenly excited. He did a quick calculation: £25,000 was roughly the equivalent of 625,000 francs. He told the others that he was almost certain that he had the answer to this question of rabbit extermination on the other side of the world. And if his remedy worked, the 625,000 francs would be his, and construction of the *Institut Pasteur* could proceed without fear of a shortfall in funding.

At this moment, Adrien Loir came hurrying into the Pasteur apartment, full of youthful enthusiasm and looking forward to seeing his cousin Jean-Baptiste. Although there was an eleven-year age gap between the two, Adrien and Jean-Baptiste had become close, firstly over the shared summers at Arbois and even more so since Louis and Marie had taken Adrien into their lives and treated him like a son. But rather than socialise, Loir found himself called on by an animated Louis Pasteur to immediately go to work.

'Adrien,' called Pasteur from his chair, 'go down to the laboratory, take a flask, and prepare virulent cultures of chicken-cholera.'[94]

Loir did not protest. With Louis Pasteur, the young man was never off duty. Loir immediately went to the laboratory and prepared the cultures as his uncle had instructed, not knowing why Pasteur needed them. When Loir joined the Pasteurs again next day, a Sunday, Pasteur instructed him to take down a letter to the editor of *Le Temps*. Loir sat at the table, took pen and paper, and wrote what his uncle dictated as Pasteur shuffled back and forth in his usual obsessive manner, aided now by a walking stick and wearing a skullcap that looked like a French gendarme's *kepi* without the peak.

'Your journal announced a few days ago,' Pasteur began, 'that the Government of New South Wales felt itself so powerless to cope with a particular pest, the increase of rabbits, that it offered a reward of 625,000 francs (£25,000) for the discovery of a means to exterminate them. Large areas of New Zealand, no less infested than Australia, are abandoned by the farmers, who have given up the raising of sheep owing to the impossibility of feeding them. Every winter the rabbits are killed in millions, but this carnage does not appear to lessen the numbers.'[95]

How did Pasteur know all this? In 1885, the New Zealand Agent-General in London, Sir Francis Dillon Bell, had written to Pasteur on behalf of the New Zealand Government, describing the extent of the rabbit plague in his colony and asking if Pasteur could suggest a way of wiping out the rabbits. Pasteur had not replied, and the following year New Zealand's Colonial Secretary, Patrick (later Sir Patrick) A. Buckley, who had been educated in Paris, had written to his London Agent-General on the subject. Buckley had wondered if it was worth trying Pasteur again, and surmised that Pasteur had not responded previously because he was in the business of saving life, not taking it. Buckley had added, in a defeatist tone, 'I presume we can give up the idea of receiving any help from M. Pasteur.'[96] It might be concluded that Pasteur had not responded to the New Zealand approach because there had not been a £25,000 carrot attached. But in his defence, 1885–86 was probably the most hectic period in Pasteur's life, a time when he was consumed with the controversial and traumatic human trials involving his rabies vaccine.

Just three months back, Pasteur had become further acquainted with the extent of the rabbit plague in both Australia and New Zealand. In late August, Marie had read her husband an article about the colonies' rabbit problem from the 15 August issue of the *Revue de Deux-Mondes*, written by Charles de Varigny. De Varigny had described how 'today, this pest is making New Zealand and Australia desolate'. Claiming that rabbits in Australia had ten litters of up to ten young at a time every year, De Varigny had said that 'they gambol about in troops' and defied the 'fruitless efforts' of the colonial landowners to destroy them. 'They are hunted, slaughtered, poisoned, and yet they swarm like ants.'[97] The article had impressed Pasteur enough for him to keep it, but not enough for him to act on it. Perhaps back in August he'd had the stirrings of the idea that now, after mention of the 625,000 franc prize, had begun to take shape in his mind on the night of 26 November.

By coincidence, only days prior to publication of the rabbit prize proclamation in *Le Temps*, Pasteur had received a letter from a New Zealand farmer beseeching him to help the landowners of his colony defeat the rabbit plague. Now Pasteur was thinking aloud as he dictated to Adrien Loir: 'Will you permit me to make known to those distant countries by the medium of *Le Temps* certain ideas, the application of which may meet with some success?'[98]

In what became more of an article than a letter, Pasteur postulated on the possibility of infecting the rabbits of the Antipodes with a disease 'which should become epidemic', one that would be communicated from one burrow to the next by infected animals so that the entire rabbit population would be wiped out. Unlike standard phosphorised rabbit poisons, Pasteur said, this rabbit remedy would be 'a poison gifted with life, like them [the rabbits], and, like them, able to multiply with a surprising fecundity'.[99]

Pasteur's mind had gone back to his chicken-cholera experiments of 1881. It had lodged in his memory that after fowls infected with chicken-cholera had died and been removed from their laboratory cages, rabbits had been placed in the same cages to await their turn in other experiments, without the cages first being disinfected. Within days, those rabbits had also died, quite unexpectedly, from no apparent cause. Subsequent autopsies showed they had been killed by chicken-cholera, and Pasteur had come to the conclusion that these rabbits had caught the disease from the chicken droppings on the cage floors. This was because he knew for a fact that in fowl yards infected chickens passed on chicken-cholera to other chickens via their droppings. 'I imagine that the same thing would happen with rabbits, and that on returning to their burrows to die they would communicate it to others who would propagate it in their turn.'[100]

As Pasteur told *Le Temps*, his past experiments had shown that it was easy to cultivate large quantities of chicken-cholera microbes in a 'liquor' of water in which meat had been boiled. To launch an infestation, he proposed to enclose a burrow with a light fence then sprinkle food that had been soaked in this liquor, or 'broth' as it came to be called, around the burrow's entrance. The rest, he proposed, would be done by the rabbits themselves once they had eaten the contaminated food. He added that chicken-cholera was harmless to farm animals other than poultry.

Pasteur concluded by saying that he had no doubts that people in the rabbit-infested countries would be ready to adopt the means he proposed, which was a very simple one, and one which was 'in every respect, worthy of a trial'.[101] After telling Loir to get the letter off to the newspaper first thing next morning, Pasteur instructed his young assistant to immediately go to the laboratory and place in a box five rabbits from the collection of experimental rabbits they kept in the basement.

At 6.00 that evening, Pasteur and Loir went to the laboratory. There, as Pasteur watched, Loir took the chicken-cholera broth he had prepared the previous evening and added a small quantity to cabbage leaves. The leaves were fed to the five hungry bunnies. At midnight, Loir returned on Pasteur's orders and added three more uninfected rabbits to the box. At 8.00 next morning, all five infected rabbits looked ill; three hours later, two were dead. By 3.00 pm, the three remaining infected animals were also dead. All five had died within twenty hours of eating the infected food. This was proof positive that chicken-cholera was fatal to rabbits, and rapidly so.

The dead rabbits were left in the box. At 7.00 that evening, one of the rabbits that had not eaten infected food was also found dead; the only conclusion to be drawn was that it had been infected by the dead rabbits. This was again good news for it supported Pasteur's hypothesis that chicken-cholera could be transmitted from one rabbit to another, even after they were dead.

The following Saturday, 3 December, the day after Jules Grévy tendered his resignation as President of France, Pasteur had Loir feed four rabbits with food tainted with chicken-cholera broth at 5.00 pm, and then place four non-infected rabbits in the same box at midnight. Next day, one rabbit from the infected batch died at 11.00 am, two at 2.00 pm, and the last at 4.00 pm. The carcasses were left in the box with the four uninfected rabbits, which continued to be fed normal food. Three of these uninfected rabbits also died over the next three days. The last rabbit died on 9 December. During this same period, Loir fed pigs, dogs, goats, sheep and rats with food infected with chicken-cholera – none showed any symptoms, and all remained perfectly healthy. Loir then repeated his rabbit experiments into the third week of December, with similar results to his first tests.[102]

In the meantime, Pasteur's letter of 27 November had been published in *Le Temps* on Tuesday 29 November and was soon picked up by other newspapers in France, Britain and elsewhere. An interested French reader of the Pasteur letter was Madame Jeanne Pommery, the seventy-six-year-old widowed owner of the famous Pommery champagne estate at Reims. On 3 December Madame Pommery wrote to Pasteur: 'Sir, I have a paddock at Reims, over my cellars, eight hectares in extent, which is entirely surrounded by walls. I conceived the foolish idea of placing rabbits there for the purpose of providing some town hunting for my grandchildren. These creatures have increased to such an extent, and undermine the grounds to such a degree, that I am anxious they will destroy them.'[103]

The widow Pommery was worried that the rabbit burrows would cave in the arches of her famous champagne cellars. Those cellars involved more than fifteen kilometres of tunnels which she'd added to caves used for storage since Roman times. The cellars were full of ageing Pommery vintages. 'If it will be convenient for you to make an experiment of the process that you recommend for the destruction of rabbits in Australia,' wrote Madame Pommery, 'I offer you an easy means of doing so.'[104]

A delighted Pasteur wrote back accepting the widow Pommery's offer; a field trial was just what the Pasteur rabbit remedy needed. By 19 December, when Loir completed his chicken-cholera experiments at the Rue d'Ulm laboratory to Pasteur's satisfaction, he was instructed to prepare to spend Christmas as the guest of Madame Pommery at her sixty-hectare estate, Clos Pompadour, at Reims. Early on 23 December, Loir took the train from Paris to Reims, 130 kilometres to the east, toting four litres of chicken-cholera broth.

Around 4.00 pm that day, Loir arrived at Clos Pompadour and went directly to the eight-hectare paddock referred to by Madame Pommery in her letter. As she had described, the paddock was surrounded on all sides by a wall. Of dry stone construction, it stood shoulder high and was topped by turf, from which grass was growing. In the distance, peeking over the top of the wall, were the towers and turrets of the eccentric Pommery château, which had been built in a variety of styles. Sure enough, the paddock was infested with rabbits – many hundreds of them. In fact, it was a rabbits' paradise for, applying a sort of Irish logic to the problem, the widow Pommery had arranged for a daily meal of lucerne and hay to be provided for the pests in the hope that this would keep them near the surface and prevent them burrowing too deep and damaging her cellars.

Loir could also see eight wooden hay wagons in the paddock, all laden with lucerne which was about to be spread around the burrows of the pampered Pommery bunnies. Just before sunset, as Pommery's workmen set to work unloading the hay, Loir pulled on a leather apron and gloves then took out his chicken-cholera jars. Once the lucerne had been distributed, he sprinkled just three bundles with the broth. He then retired to the Clos Pompadour château which, incidentally, had been occupied by the Prussian governor of Reims during the Franco-Prussian War. There at the château Loir would spend Christmas with the Pommery family, enjoying their hospitality, their food and their champagne: the expensive Pommery brand was a favourite of aristocratic Europeans and rich Americans.

On the Saturday morning, the day after his arrival at the Pommery estate, Loir went out early with one of the estate workmen to survey the rabbit paddock. They counted nineteen dead rabbits lying outside their burrows. Of the many other rabbits that made the paddock their home, there was no sign. A visitor now arrived; Monsieur Trompette, a commercial photographer from Reims, had been alerted that Louis Pasteur's nephew was conducting biological experiments with the widow Pommery's rabbits. As Monsieur Trompette set up his tripod and camera, Loir explained that rabbits generally retreated into their burrows to die, and that many other carcasses could be expected to litter the paddock's burrows. The photographer then had Loir and the middle-aged workman pose in the act of studying a dead rabbit and snapped them for posterity.[105]

The next day, a Sunday, no inspection was made as Loir enjoyed a relaxing day with his hosts. Overnight there was a light fall of snow, so when Loir made an inspection on Monday morning he was able to observe that, in addition to thirteen fresh corpses lying on the ground, there were no paw prints in the snow; there had been no rabbit activity overnight. Lucerne was distributed as usual that day; by evening, it had not been touched and not a single rabbit was seen. Madame Pommery, meanwhile, showed Loir several English newspapers that had recently arrived; all of them contained articles critical of the chicken-cholera rabbit remedy that Pasteur had proposed in his letter to *Le Temps*.

A lengthy, sarcastic editorial in the *Times* of London was typical of the English response to Pasteur's proposal: 'The rabbit race must be prepared to encounter a formidable antagonist to its progress in the annexation of Australasia,' the unnamed editorial writer began, tongue firmly in cheek. Pasteur, it scoffed, 'concocts in his chemical kitchen a microbe soup which simply has to be sprinkled over vegetable bait in order to produce the desired catastrophe among the rabbit tasters'. Taking the side of the rabbit, the *Times* felt the Pasteur tactic underhanded, and inhuman. It suggested that the British weasel and stoat should be introduced into Australia and New Zealand to solve the colonial rabbit problem. That, according to the paper, would be a much more sporting solution.[106]

The formidable Madame Pommery, who had been running the Pommery business on her own for decades, was incensed by the attitude of the English press. She pointed to her previously overrun paddock. 'What becomes of the English critiques in the face of such a result?' she asked. 'A paddock of eight hectares, swarming with rabbits, has become a field of death. Monsieur Pasteur poisons one ordinary meal for the rabbits and a few days afterwards nothing lives. Every one is put an end to. Every one is dead!'[107]

On the Tuesday, Loir set off for Paris. There could be no doubting now that a small amount of chicken-cholera effectively killed rabbits, and killed them quickly. Most importantly, Loir had proven that chicken-cholera worked just as well in the open as it did in the laboratory. News of Loir's accomplishments at the Pommery estate was soon appearing in the French and foreign press. 'The scheme for the destruction of rabbits which M. Pasteur lately recommended to the farmers of New South Wales has, it is said, been tried with success on Madame Pommery's property near Reims,' said

London's *Daily Telegraph*, in a report picked up and reprinted by the *New York Times*.[108]

In the first week of January, Madame Pommery wrote to Pasteur to inform him that, according to her workmen, in excess of a thousand rabbits had previously enjoyed the free meals they had doled out in the paddock above the Pommery cellars. Now, said the delighted *grande dame*, the paddock was heaped with carcasses; not a solitary rabbit had survived.[109]

Loir's report, when he arrived back at the Rue d'Ulm, excited Pasteur, who began making plans at once. Loir, he said, would go to New South Wales to demonstrate the effectiveness of the chicken-cholera rabbit remedy and then collect the 625,000 franc prize. Loir would need an assistant – ENS student Louis Momont, Emile Roux's cousin, immediately came to mind.[110] How soon could the pair leave? The sooner the better. Pasteur instructed Loir to urgently obtain information on sailings to Australia by British steamship lines.

Pasteur then decided that he had better write to the New South Wales Government to let them know that he was serious about claiming the rabbit destruction prize and was sending his representatives to demonstrate the efficacy of his rabbit eradication scheme. He had already been in touch with the British Embassy in Paris, and it had informed him that New South Wales had an agent-general, a sort of ambassador to Europe, based in London. His name was Sir Saul Samuel.

On 5 January, Pasteur dictated a letter of many pages to Loir, addressed to Agent-General Samuel. Headed 'Upon the destruction of rabbits in Australia and New Zealand', the letter began by quoting in detail from Charles de Varigny's August *Revue de Deux-Mondes* article, before citing Frank Abigail's rabbit competition proclamation and repeating, word for word, Pasteur's 27 November letter to *Le Temps*. Pasteur then described Loir's chicken-cholera experiments in the laboratory between 27 November and 18 December, and the subsequent field trial at Madame Pommery's estate, as well as quoting from several letters from Madame Pommery herself. For some reason, he also enclosed a copy of the deprecatory London *Times* article, which Madame Pommery had sent on to him.[111]

'In concluding this letter,' Pasteur said, 'I have the honour to state, for the information of the Agent-General, that it is my intention to dispatch two experts to Australia and New Zealand for the purpose of putting into practice in those remote countries the process that I have just explained. Finally, I express the hope that my scheme may be permitted to compete for the reward of £25,000 sterling that has been lately offered by the Government of New South Wales for the extermination of rabbits in Australia.' After his signature, Pasteur noted that he was a member of the French Institute, of the Royal Society of London, and of the Royal Society of New South Wales – he had been granted membership of the New South Wales society several years before.[112]

Shortly after sending this letter to London, Pasteur learned that steamers of Britain's Orient Line left England for Australian ports every two weeks, and that one would be stopping at Naples in late February on its way to the Suez Canal and Australia. Pasteur decided that Loir and his assistant would take the train to Marseilles, then board a French ship to cross the Mediterranean from Marseilles to Naples, where they would join the Orient Line steamer. Consulting the sailing schedules, Pasteur worked out that his representatives should be in Naples for the Orient Line departure from that port on 28 February. A French ship sailing from Marseilles on 8 February would put Loir in Naples in plenty of time; he would have to spend a week or two in an inexpensive hotel in Naples until the British steamer arrived, but that was acceptable.

It was decided: Loir would sail for Australia on 8 February. Two First Class tickets for the 8 February sailing from Marseilles and for the 28 February sailing from Naples were promptly purchased from a Paris shipping agent. As usual, Pasteur was impatient to move forward – in this case impatient to send his team to Australia, demonstrate his method, and claim the £25,000 prize. Over the years, Pasteur biographers have praised what they have perceived as his tireless patience, for spending years working on one research project or another before finding the answer he was looking for. This was not patience; this was stubbornness, and obsessiveness. As Pasteur had demonstrated time and again with his rush to trial new vaccines, often against the advice of his more cautious associates, he was in fact incurably impatient.

On 8 January, someone – perhaps Loir or perhaps Marie Pasteur – reminded Pasteur that the 9 November New South Wales Government advertisement in *Le Temps*, which had been repeated in the paper's 2 December edition, had specified that all communications relating to the competition be directed to Secretary for Mines Francis Abigail in Sydney. Pasteur had a sudden panic. Fearing that the letter to the Agent-General in London might be disregarded and that the Pasteur competition entry might not be considered, he now composed a hasty letter to Abigail himself: 'Sir, I have the honour to submit this communication, in reply to the official notification published by your direction in *Le Temps* of Paris.' Pasteur even attached a cutting of Abigail's advertisement. 'The delegates whom I am sending to Australia will leave Marseilles for Melbourne on the 8th of February next. With the hope that they will be accorded a hearty welcome in Australia, and that my scheme will answer all expectations in connection with the destruction of the rabbits that are devastating your fair land, I have the honour to request your acceptance of the assurance of my sincere regard.'[113]

In his 5 January letter to the Agent-General, Pasteur had given his address as '45 Rue d'Ulm, Paris'. On this letter to Mines Secretary Frank Abigail three days later, he gave his address as 'Pasteur Institute, Paris'. In Pasteur's mind, his institute was already a reality, courtesy of the New South Wales rabbit prize, which he felt was as good as his.

The letter to Abigail was put in the post. Eleven days went by. Eleven days during which Pasteur began to increasingly fret that he had not heard anything from either the New South Wales Government or Agent-General Samuel in London. He could not know that Sir Saul Samuel was at that time on his way from England to Australia for a visit home, and that an Acting Agent-General, Sir Daniel Cooper, was about to fill in for him in his absence. Loir, meanwhile, would have told his uncle that his letter to Australia would take at least a month to reach Mr Abigail, and that it would take another month for a reply to come back to Paris. But Pasteur felt sure that the contents of his letter in *Le Temps* on 29 November must have been communicated to the New South Wales Government by this time; the Australians must be aware of his interest in their rabbit eradication competition.

A reason for the silence was now suggested to Pasteur: perhaps his presumptive declaration that he was sending representatives to Australia, uninvited, may have offended the New South Wales Government? Perhaps they did not appreciate his lack of courtesy, and had chosen to ignore him? So on 19 January, Pasteur sent a telegram to Australia. It read: 'Abigail, Secretary, Mines, Sydney. Have you received the communication which appeared in the *Temps* newspaper? I have since made conclusive experiments. May I send immediately delegates with the certainty that the Board will investigate the process? L. Pasteur.'[114]

The cost of this telegram to the other side of the world would have been enough to give the frugal Pasteur another stroke – at 9 francs a word, it amounted to 342 francs. This was the equivalent of one third of the Emperor of Brazil's entire donation to the Pasteur Institute subscription fund. It is likely that Pasteur had never before in his life sent a single telegram; anyone who had experience with telegrams knew to cut out unnecessary words to save money. When he overcame the shock of the cost, Pasteur would have vowed to never again send a telegram to Australia; or to anywhere else for that matter.[115]

But at least his expensive telegram bore fruit. Two days later, a brief and cost-conscious response came clicking through the telegraph cables that ran under the oceans of the world and was delivered to the Rue d'Ulm by a telegram boy. The message was from Sir Henry Parkes, the Premier of New South Wales, and had been directed to Acting Agent-General Sir Daniel Cooper in London. The French-speaking Cooper, who had lived and worked in Le Havre as a young man, had retransmitted the telegram in French to Pasteur in Paris. Parkes' telegram said: 'Kindly ask Pasteur to let you have the chicken-cholera germ and tell you how to use it.'[116]

Pasteur must have been ecstatic. Here was the head of the New South Wales Government expressing an unequivocal desire to use

Pasteur chicken-cholera in New South Wales. The fact that Parkes wanted Pasteur to hand over his microbes to the Agent-General was beside the point – that was out of the question; it was like a patient telling a surgeon that he would carry out his own amputation. Pasteur would never allow anyone but himself or his assistants to handle his microbes, both for proprietary reasons and for reasons of safety. But he could, he was sure, soon sort this out with the New South Wales Government. The fact remained, here was proof that the Government of New South Wales knew about Pasteur's rabbit eradication method, and wanted it.

This was confirmed to him when, in the last week of January, Sir Daniel Cooper sent young Dr Frank Hinds from London with a letter recommending Hinds' inclusion, as an English speaker, in the team of delegates that Pasteur was determined to send to Australia. As Loir and Germont prepared for their departure from Paris for Marseilles in the first week of February, Pasteur would put Hinds through a ten-day crash course in the cultivation of the chicken-cholera bacillus and how to use it to exterminate rabbits. But first Pasteur would pay a visit to his bankers, taking with him the telegram from Sir Henry Parkes as written proof that his rabbit competition entry was sought after by no less than the Prime Minster of New South Wales.

Pasteur could now tell the Paris bankers that, within a matter of months, his representatives would without doubt be returning with the 625,000 franc prize. In his scientific opinion, no other entrant could possibly devise a method superior to his. And had he not proven his method, in the laboratory and in the field at Madame Pommery's estate? All his delegates had to do was demonstrate the astounding effectiveness of his method and come home with the prize. As far as the construction of the new Pasteur Institute was concerned, Pasteur might have to make one or two economies to trim the budget a little, but with the 625,000 francs from Australia, to his mind there was absolutely no reason why construction could not begin at once.

To Pasteur's joy, the bankers agreed with him. By the time that Adrien Loir, François Germont and Frank Hinds sailed from

Marseilles on 8 February on the first leg of their journey to Australia, taking two small boxes containing chicken-cholera broth with them, full-scale construction of the Pasteur Institute was under way at the Grenelle site.

8

BUREAUCRATIC WARFARE

In the New South Wales of the 1880s, just as governments could change in the blink of an eye, so plans and policies were apt to take a U-turn overnight. Such was the case when Louis Pasteur's bid to win the rabbit eradication prize met its first hurdle in New South Wales almost before it got off the ground.

By December 1887, Frank Abigail's international newspaper advertisements had generated a torrent of rabbit competition submissions from around the Australasian colonies and from around the world. As bulging mail bags filled his office, Abigail became increasingly confident that his competition would result in the total eradication of the rabbit plague in the colony. But the squatters of New South Wales did not share his confidence. When Abigail and Premier Parkes had announced the imminent end of the rabbit bounty system, both the rabbiters and the squatters were up in arms. Rabbiters complained that their livelihood was being taken from them while the squatters said that they could not and would not bear the exorbitant burden of rabbit extermination on their properties.

When Abigail had held his ground against the squatters, declaring that courtesy of his competition there would soon be no rabbit problem to worry about, the squatters petitioned the government to reduce the rents they were paying on their leasehold properties. If they had to carry the cost of rabbit extermination, they said, having no faith in Abigail's competition, then the government would have to help them another way, via rent reduction. Many newspapers took the same view as the influential squatters and some even put a case for the rabbiters' continued employment at government expense to prevent a rise in unemployment numbers in the colony – Sydney's jobless men were said to total 10,000 in 1887 according to one of Parkes' political opponents.[117] When Abigail would not bend on rents, the squatters lobbied other members of the Parkes' cabinet behind his back.

In a cabinet meeting in December, Frank Abigail was ambushed by cabinet colleagues. Land Minister Thomas Garrett, expressing the view that the rabbit eradication competition had yet to generate Abigail's hoped-for remedy, and may never do so, proposed that the squatters' request for reduced rents be agreed to. And a majority of the ten-member cabinet voted with him. Although Sir Henry Parkes supported Abigail, the Secretary for Mines was 'rolled' by his colleagues and the motion was carried. Humiliated, Abigail, who had previously declared that nothing nor no one would make him back down on the matter of rent reduction, threatened to resign. Behind closed doors, Premier Parkes talked his fiery young colleague out of giving up his portfolio by, typically, devising a way for him to save face.

Parkes transferred the Rabbit Department from the control of the Secretary for Mines to the control of the Minister for Lands. It would now be Thomas Garrett who brought in the rent reductions. He would also extend a reduced form of the rabbit bounty system into 1888 while at the same time taking responsibility for the £400,000 budget deficit being carried by the Rabbit Department. While Abigail would no longer have to lose sleep over rabbits, he lost control of the rabbit competition that he'd launched with such enthusiasm. With the Rabbit Department no longer under Abigail's control, the competition lost its first champion.

Although Abigail had been sidelined, Premier Parkes continued to support the competition. By the end of the third week of January,

Sir Henry had received a package from the Agent-General's office in London containing a translated copy of Pasteur's *Le Temps* letter. The expensive telegram from Pasteur to Frank Abigail, in which Pasteur asked permission to send his representatives to Australia, was also passed onto Parkes. These two communications prompted the Premier's telegram to Sir Daniel Cooper asking for him to approach Pasteur to obtain his chicken-cholera microbes. Before long, Cooper, a one-time Free-trade parliamentarian and former Speaker of the New South Wales Legislative Assembly who was Parkes' loyal friend, had cabled the Premier that Pasteur would not part with his microbes.

At this time Parkes was unaware that Pasteur had already booked his representatives' passage to Australia. At this time too, a member of Parkes' cabinet and another close friend, Julian Salomons – later Sir Julian, a former Chief Justice and now the government representative in the Legislative Council – was in England on a trip intended to improve his flagging health. Parkes cabled Salomons, a fluent French-speaker, and asked him to visit Monsieur Pasteur in Paris to see whether he could convince him to allow Salomons to bring his microbes back to Australia with him for testing when Salomons set off for New South Wales in March.

Meanwhile, Parkes passed on Pasteur's *Le Temps* letter and the Abigail telegram to Thomas Garrett who, as Lands Minister, was now in charge of the rabbit competition. Garrett had already sifted through many hundreds of competition entries and kept aside several that he felt best met the conditions set out by the proclamation advertised around the world. One of these was a joint local submission from Drs Henry Ellis and Herbert Butcher. Dr Ellis was a thirty-one-year-old Irish-born physician located at Double Bay in Sydney's eastern suburbs while Dr Butcher, a much older man, lived at Wilcannia and had a financial interest in nearby Tintinallogy Station. The Ellis–Butcher submission revolved around an as yet undiagnosed disease that had recently begun to kill large numbers of rabbits on Tintinallogy Station.

Another entry of note was from Dr Archibald Watson, Professor of Anatomy at Adelaide University in South Australia. Watson believed that he had discovered a fatal disease in rabbits, so-called rabbit scab. Watson was actively selling rabbits infected with 'Watson's disease' to South Australian farmers with the assurance that these animals would infect and eradicate other rabbits on their properties. Both Watson's submission and that of Ellis and Butcher were accompanied by wads of supporting material in the form of extracts from newspaper articles, letters from government officials and details of the entrants' own experiments and those of New South Wales Government Veterinarian, Edward Stanley, involving rabbits and the entrants' particular diseases. Pasteur's entry, by comparison, was sadly lacking in detail.

On Wednesday 8 February, the same day that Loir, Germont and Hinds set sail from Marseilles for Australia, Minister Garrett bundled up the Ellis–Butcher submission, the Watson submission, and Pasteur's *Le Temps* letter and 19 January telegram, and sent them all to the Premier with the request that he submit the material to the Board of Health, which came under Parkes' direct control via the Colonial Secretary's Department, 'for their consideration and report as to whether the introduction of these diseases, as proposed, would be dangerous to human life or health'.[118]

The pragmatic Parkes would have much preferred to have simply acquired the microbes and done away with the rabbits without any involvement from the colony's bureaucracy. He also knew that Garrett was not fond of the rabbit competition and would be quite happy to see the whole thing filed away under 'impractical'. Now that Garrett had quite deliberately put his request for Health Board involvement on record, Parkes could not be seen to be taking risks with the public health. So that same day the Premier forwarded on the documentation, with his hand-written note on Garrett's letter, 'Referred for report of the Board as desired by the Minister for Lands.'[119]

The file then disappeared, to turn up on the desk of the Treasurer, John Fitzgerald Burns, five days later. Treasurer Burns sent the file to the Board of Health on Monday 13 February. It may or may not be coincidental that, in the meantime, on Thursday 9 February, Dr W. Camac Wilkinson, a Free-trade-supporting member of the

Legislative Council, wrote a long and verbose letter to Minister for Lands, Garrett. This was the day after the Premier had onforwarded the rabbit competition material for the Health Board's advice. It is quite possible that, at Lands Minister Garrett's prompting, Treasurer Burns deliberately held back the material while Garrett tipped off his friend and their fellow MP Wilkinson that Parkes was sending it to the Board of Health at Garrett's request, giving the doctor a chance to intervene.

Dr Wilkinson's letter began with the remark that 'there seems to be a general idea that, of all the methods proposed for the extermination of the rabbits, Pasteur's method is the one which promises best'. But, he said, 'it is right to pause and consider'. Wilkinson expressed grave concerns about Pasteur's proposal to introduce chicken-cholera into New South Wales. 'Pasteur now, with every good intention, and in the best of faith, is offering us a living weapon with which to combat the rabbit pest,' Wilkinson wrote. 'I am not in the mind to put my whole trust in M. Pasteur's sanguine anticipations that we shall have all good and no evil results.'[120]

He went on to belittle Pasteur's work claiming, incorrectly, that his anthrax findings had been largely disproved since his work with that bacillus. And then he asked, 'Can we say that the virtues, which he so positively claimed for vaccination with his attenuated virus, are of any practical value?' Therefore, he said, until they were in a position to know from their own investigations that fowl cholera would destroy rabbits without doing any other harm, 'we should positively forbid the introduction of this disease into the Colonies'.

There was a possibility, Wilkinson continued, that some desperate landowner might secretly introduce chicken-cholera into the colony themselves to combat the rabbit plague. To prevent this, he recommended that Pasteur's method 'should have a full and fair trial' on an island somewhere. But the trial should be conducted by local experts, he stressed, not by visitors from Europe.[121] Wilkinson ended his letter by declaring that the government should encourage scientific work in the colony. 'Such encouragement is given preeminently in Germany, and Koch's splendid work has in part flown from it.' That encouragement, he told Garrett, should take the form of 'a

Bacteriological Laboratory, either at the University or in connection with the Board of Health, as I urged three years ago'.

On 10 February, Lands Minister Garrett passed Dr Wilkinson's letter on to the Premier's office, asking that it be forwarded to the Board of Health to be considered along with the rabbit competition material that had previously been sent to them.[122] The Premier duly passed the letter onto the Board of Health.

On 22 February, Edward Sager, Secretary of the Board of Health, delivered a short report from his office at 127 Macquarie Street two doors along from the Premier's office in the Colonial Secretary's Department. Having considered the material submitted by the Minister for Lands and the letter from Dr Wilkinson, the Health Board, Sager wrote, was of the opinion that 'it would be too much to expect, as some writers seem to do, that the introduction of any infection would be so successful as to cause complete extermination of the rabbits'.[123]

It was clearly impossible to know, Sager's letter went on, whether the Ellis–Butcher or Watson diseases might not be communicated to other animals or to humans. As for the disease recommended by Monsieur Pasteur for eradicating the rabbits, the report writer referred to Pasteur's *Le Temps* article and scoffed that, 'he *imagines* that it could be readily propagated among them with very fatal results'. The report then noted that Pasteur had made no reference to the possibility of the communication of chicken-cholera to humans.[124] Pasteur had not mentioned this because there was no case on record anywhere of humans contracting chicken-cholera. Being eminent medical men, the members of the Medical Board would have known this too but the report's author cynically used Pasteur's sin of omission against him.

The report summed up that the government would be taking a grave risk in introducing a measure, the results of which were at that time unforeseen, which might turn out to be harmful to human beings or to stock. Once chicken-cholera had been introduced it said, 'it would be beyond the power of any authority' to undo the harm that was done. 'The Board would therefore recommend that the measure proposed for the introduction of infection among rabbits should not be adopted.'[125]

What Sir Henry Parkes said after he received this Health Board report is not recorded. He was not noted for having a temper but he would have been far from pleased by the patronising and dismissive tone of the report and its unsupported conclusions. Certainly, volatile Frank Abigail would have exploded when he heard about the report. The gentlemen of the Board of Health, he would have declared, were idiots. How could they have ruled against the use of any disease to destroy the rabbit population with only very limited information available, and without any scientific tests yet having been conducted in the colony? The government had never proposed to simply introduce some rare disease into the country, as the report implied. The rabbit competition's rules, as drawn up by Abigail, specified that selected methods must be tested for a full year before any decision was made on implementation of one scheme or another.

For the moment, Premier Parkes did nothing in response to this negative report because he knew, via a telegram from Sir Daniel Cooper in London, that more written material from Pasteur was due to reach Sydney at any time.

Three days later, on Saturday 25 February – all government employees then worked six days a week – the permanent head of the Lands Department, Under Secretary Charles Oliver, who was Minister Garrett's deputy, received several letters from Louis Pasteur which had just arrived by steamer from Europe. There was Pasteur's correspondence of 5 January to the Agent-General in London, with its quotations from the Charles de Varigny's newspaper article; his lengthy discussion of Adrien Loir's laboratory experiments with chicken-cholera and his field trial at the Pommery estate; and the attached editorial from the London *Times*.

The second letter Oliver received was the brief one that Pasteur had sent to Frank Abigail on 8 January informing him that his delegates were leaving Marseilles exactly one month later, bound for Australia. Oliver placed the letters on Garrett's desk, ready for the minister's attention as soon as he came into the office on Monday, with a note suggesting that perhaps the Premier would want these letters to be placed before the Board of Health to see whether the board would now want to modify its opinion about the introduction of chicken-cholera.

The next day, across the world in Paris, Louis Pasteur received a Sunday visit from the daughter of Acting Agent-General Sir Daniel Cooper who, on behalf of her father, asked if Pasteur would be so kind as to allow Julian Salomons and Sir Francis Dillon Bell, the New Zealand Agent-General, to visit him. Pasteur of course agreed. Miss Cooper also passed on the information from Sir Daniel that in Australia there was a fatal disease in cattle and possibly also sheep called, in the colonies, Cumberland disease. Cooper was wondering if this was in fact anthrax and if so, whether Pasteur's vaccine might be used in Australia. From Miss Cooper's description, Pasteur believed that this Cumberland disease did indeed sound very much like anthrax.

After the Agent-General's daughter left the Rue d'Ulm, Pasteur began dictating a long letter to his wife who now, in Adrien Loir's absence, was serving as her husband's secretary. Addressing the letter to Loir care of Sir Henry Parkes in Sydney, Pasteur told his nephew about Miss Cooper's visit and about their discussion regarding Cumberland disease. 'If it is anthrax,' Pasteur wrote to Loir, 'I will consider vaccination. If it is another microbe, send me a tube of blood.'[126] He held off sending the letter until after the visit from the two gentlemen from the Antipodes, who might have some interesting comments to contribute.

The following Monday, 27 February, Minister Garrett came into his Sydney office. Finding Pasteur's letters waiting for him, Garrett asked Charles Oliver to indeed pass them on to the Premier, with the hope that he would send them on to the Health Board. The letters reached Parkes later that day. He immediately forwarded them on to the Health Board. Board Secretary Sager received them the next morning, and put them on the agenda for the Health Board's scheduled meeting the next day, 29 February.

Parkes cabinet member Julian Salomons met with Louis Pasteur at the Rue d'Ulm around this time. Both Salomons and Sir Francis Dillon Bell had lived in France in their youth so both could speak French like natives. Salomons was also a great wit and excellent company so he chatted easily with the scientist and put him at his ease. When Salomons passed on Sir Henry Parkes' request that he be provided with the chicken-cholera microbes, Pasteur informed his guest that his delegates were well on their way to Australia with his microbes, having left Naples several days earlier.

Undaunted by this information, Salomons showed great interest in the chicken-cholera method of rabbit destruction. In response, Pasteur became animated and took his visitor down to his laboratory. There Pasteur and a student-assistant conducted various experiments with the chicken-cholera bacillus for Salomons' benefit, with Pasteur explaining how his microbe soup would solve all the Australian and New Zealand rabbit problems. As Salomons was leaving, Pasteur, having learned that his visitor was not only a member of the New South Wales cabinet but a Queens Counsel, the most senior form of lawyer in the British Empire, asked Salomons if he might not act as his 'friend in court' with the New South Wales Government to ensure he received fair play. Salomons departed assuring Pasteur that he would have no reason to suspect that he would receive anything but fair play from Salomons' good friend Sir Henry Parkes.

A day later, New Zealand Agent-General Sir Francis Dillon Bell paid Pasteur a visit. Bell, an immensely wealthy man, told Pasteur during a long chat that he had massive landholdings in New Zealand which were managed by one of his sons. But, said Bell, between the rabbit plague and the declining wool price he was finding his mortgages a heavy burden to bear. Pasteur gave Bell a similar laboratory tour to that which Salomons had experienced. And like Salomons, Bell went away convinced that Pasteur's rabbit eradication method was the solution to the colonies' rabbit problems.[127]

At the same time, Pasteur's eyes were opened by what Bell had told him. When, after Bell's departure, Pasteur resumed dictating his interrupted letter to Adrien Loir, he informed his nephew that he'd just been visited by several colonials. 'One of them told me that he had a 3,000,000-hectare property and 1,000,000 sheep. It is frightening! I abandon the idea of a lethal culture to be distributed here and there on these large areas.' Now, he told Loir, he was thinking in terms of establishing in Sydney 'a single factory that will prepare the culture to be sent to farmers or collected by them. It will be easy'.[128]

Salomons was back in London two weeks later, preparing to head for Plymouth to catch an Orient Line steamer for the trip home to Sydney. In London, he was approached by the English correspondent of the Melbourne *Age*. The reporter had been speaking with Sir Francis Dillion Bell from whom he'd learned that the pair had recently visited Louis Pasteur to discuss rabbits. 'The good scientist has convinced them both that his scheme is thoroughly feasible and, what is of more importance, safe also,' reported the *Age* man. 'After carefully witnessing numerous experiments' Salomons 'became so enthusiastic a convert that he laughingly declared his intention of dining off a rabbit which had become a victim of "microbe soup"'.[129]

Nobody was laughing in Sydney. The Board of Health met on 29 February and immediately sent a curt reply to the request from the Premier that it reconsider its decision of 22 February in light of the two letters recently received from Louis Pasteur. 'The Board, having taken these papers into their careful consideration, see no reason to modify their opinions expressed in their resolution of the 22nd instant.'[130]

The very next day, 1 March, Sir Henry Parkes acted. He wanted to see Pasteur's science employed in New South Wales in the same way that it was in France and other parts of the world, and he was not going to let pedantic politicians or obstructionist Health Board members stand in the way of progress. He summoned Lands Minister Thomas Garrett to his office. And there he informed Garrett that over the past fourteen months he had been greatly disappointed that Garrett's other interests had prevented him from attending many cabinet meetings. Since Mr Garrett was not able to devote the necessary time to his government duties, Sir Henry would be forced to accept his resignation from the post of Minister for Lands. Effective immediately. A stunned Garrett returned to his own office and wrote and signed his resignation.

Parkes now took personal charge of the Rabbit Department and the rabbit eradication competition by appointing himself, in addition to all his other roles, Acting Minister for Lands. He did not stop there. The Board of Health's two rulings, which had been widely covered in the local press, were troubling but not insurmountable. Parkes would not attempt to personally overrule his Board of Health, a move sure to give his political opponents ready ammunition. He would astutely sidestep it.

For two decades, Henry Parkes had spoken about the wisdom of the eventual federation of all the Australasian colonies into one nation but he had never taken any concrete steps in that direction. This was soon to change. In a little over a year's time, Parkes would make a speech at the New South Wales town of Tenterfield, known today as the Tenterfield Oration, in which he would actively promote the federation of the colonies. He would go further. Following the address, Parkes would invite all the premiers of the other colonies to a Federal Convention where they would begin the process of mapping out a constitution for a new nation of Australia. For this initiative, which would see the colonies travelling down the road to eventual federation in 1901, Parkes would go down in Australian history as the 'Father of Federation'.

But that road had begun prior to Tenterfield, and it was all to do with rabbits. On 1 March 1888, after dismissing Thomas Garrett and taking control of the Rabbit Department and of the rabbit destruction competition, Parkes cabled all the premiers of the other colonies of Australia and New Zealand. He knew that Louis Pasteur's delegates had left Marseilles on 8 February and, from Julian Salomons via Sir Daniel Cooper, he had learned that they had left Naples on 26 February and would therefore reach Sydney by early April. Now Premier Parkes invited his fellow premiers to send delegates to Sydney who would, from the middle of April, sit on an intercolonial commission that would decide on the winner of the rabbit eradication competition inaugurated by New South Wales. This unique collaboration between all the rabbit-plagued colonies could contribute to and benefit from the outcome of the competition.

If Parkes' fellow premiers accepted his invitation, this would be the first time that the Australasian colonies had collaborated in a common cause. The road to Australian federation thus began in Sir Henry Parkes' office in March 1888 – prompted by rabbits and the science of Louis Pasteur.

9

THE BREWER'S OFFER

ADRIEN LOIR, HIS FELLOW Frenchman François Germont, and their dud interpreter Frank Hinds, found in Australia a very different world to the one they had left behind in Europe. The differences extended beyond the unique flora and fauna and the fact that here the seasons were the opposite to those of the northern hemisphere. In Australia customs, attitudes and interests were quite unfamiliar to the visitors. As Adrien Loir ventured into the fifty-three-year-old city of Melbourne he found that the locals were little interested in the affairs beyond Australian shores. They were more concerned with celebrating the centenary of white settlement of Australia, were talking about countering the wave of Chinese immigrants washing into the colonies from southern China and apparently threatening their jobs, or were preparing for the upcoming season of the rough-and-tumble game of Australian Rules football – footer or footy these wild colonial boys called it. To visitors it seemed an excuse for assault and battery.

Downtown Melbourne was bustling. Imposing public edifices of grim grey stone lined the inner-city streets, a counterbalance of broad avenues and narrow lanes. Trams rattled and clanged through the crowded thoroughfares, which were filled with horse-drawn omnibuses, hackney cabs, carriages of the well-to-do and wagons and drays of every description, all nose to tail throughout the commercial district. Traffic jams were not unusual for there were no traffic signals or traffic police. Here, there were no sidewalk cafés; there was the occasional coffee house, but this was a land of tea drinkers. Wine drinkers were even rarer. Beer and spirits were the colonial beverages of choice.

Within days of arriving in Melbourne, Loir was approached at the club where he was staying by a local brewer, Thomas Aitken, whose Victoria Brewery had been producing Victoria Bitter, or VB as its many consumers had been calling it since 1854. A canny Scot, Aitken had read all about the great financial savings that the Danish brewers of Carlsberg Beer had made by adopting Pasteur's techniques. Aitken had been planning to travel to Europe to learn the Pasteur secret so he could apply it in the VB brewery but here was Pasteur's nephew and assistant right on his doorstep.

Aitken asked Loir, 'Could you show me how to culture brewer's yeast?'[131]

Of course Loir knew everything there was to know about culturing yeast, just as he was familiar with all Pasteur's research and techniques. As it happened, Pasteur had sent Loir to Copenhagen in 1886 to initiate the new chief of the laboratory at the Jacobsen Brewery into the latest Pasteur yeast-culturing techniques for their Carlsberg brew. The Jacobsen laboratory had in fact gone on to develop a range of cultured yeasts, each producing a different flavoured beer. Loir took immediate interest in Aitken's proposal.

By this time, Frank Hinds had arrived in Sydney where he took up residence at the city's Union Club and called at the office of Mines Secretary Abigail to ascertain when his French colleagues might be allowed to come to Sydney with their chicken-cholera microbes. After being redirected to the office of Sir Henry Parkes, Hinds wrote to Loir in Melbourne telling him that in response to his approach the New South Wales Government had informed him that they were 'examining the question'.[132] Loir had no idea how long he might have to wait in Melbourne for a decision from the New South Wales authorities, and he was worrying about running out of money and being forced to board a ship and use his return ticket to go home. When brewer Thomas Aitken offered to pay Loir for showing him how to culture yeast, the young Frenchman snapped up the offer.

For the next week or so, Loir went to work at the Victoria Brewery's East Melbourne plant, teaching Aitken culturing techniques with the aid of his trusty microscope. Aitken was thrilled with the results. When the task was completed, he thanked Loir profusely and handed him £250. Loir was stunned. This was more than 6000 francs, $100,000 in today's currency. The young scientist had never seen so much money in all his life.

But Aitken was more than pleased with the deal. 'You've allowed me to save £3000,' the brewer assured Loir. 'I was going to travel to Europe to learn all about these methods.'[133]

Loir was also relieved; Aitken's payment took the financial pressure off the Pasteur party. 'We now had money, we could stay in Australia,' Loir was to write.[134] As for Victoria Bitter, it went on to outlast and outsell numerous competitors and become the top-selling beer in Australia. Today a part of the Carlton and United Breweries stable, VB accounts for one-third of all packaged beer sales in the country –[135] thanks in part to a little help from the affable Adrien Loir and the science of Louis Pasteur.

On the afternoon of 17 April, Loir and Germont quickly packed their bags, checked out of the club that had been their home for two weeks, and hurried to Flinders Street Station to catch the Sydney Express. Loir had just received a telegram from Hinds, urging him to bring the microbes to Sydney at once.

10

THE RABBIT COMMISSION MEETS

AT 10.00 IN THE morning of Monday 16 April 1888, the Intercolonial Royal Commission of Inquiry into Schemes for the Extermination of Rabbits in Australasia met for the first time in Sydney. Sir Henry Parkes' call to his fellow leaders of the Australasian colonies had yielded results. New South Wales had sent three delegates to the Commission. Victoria, not to be outdone, had also nominated three delegates. South Australia had sent two. Queensland had sent one, but would later add a second. Tasmania had chosen one delegate, and New Zealand had also nominated one. The British Governor of Western Australia – which would only become a self-governing colony in 1890 and was at this time controlled from London – had declined to send a delegate, making Western Australia the only Australasian colony not represented.

This body, called simply the Rabbit Commission by the colonies' newspapers, was to be, in effect, the judging panel for the international rabbit eradication competition. On its recommendation, Sir Henry Parkes would award the £25,000 prize. Over the coming weeks and months, the personalities and private agendas of the men sitting on this Commission would determine how Louis Pasteur's science would be accepted and used in Australia. So, to appreciate the difficult path that Adrien Loir would have to tread on his uncle's behalf, it is important to know a little about those personalities and their agendas.

This first morning, ten men – nine delegates and a Secretary to the Commission – sat at a long table at the eastern end of the Executive Council Chamber on the third level of the Colonial Secretary's Department, just a short distance along the corridor from the Premier's office.[136] The delegates knew they may be required to devote an amount of their time to the Commission for up to twelve months, while their governments were aware that they would be responsible for their own delegates' expenses for travel, accommodation, and food during that period. New South Wales, as the colony most affected by the rabbit plague and as the creator of the competition, would fund the prize payment.

Leading the New South Wales delegation was fifty-one-year-old Dr H. (Henry) Normand MacLaurin. Born in Scotland, a graduate of Edinburgh University, MacLaurin had been a ship's surgeon with the Royal Navy before settling in New South Wales and taking various medical and university appointments. By this time he was Vice Chancellor of Sydney University. Since 1885 he had also been Government Medical Adviser and, more poignantly, Chairman of the Board of Health; both appointments had been made by a Protectionist Government, for MacLaurin was close to the Protectionist leaders – Sir Henry Parkes' political opponents. It had been MacLaurin who'd authored the two Health Board reports which in February had recommended that, in the interest of safety, permission should not be given for Pasteur's chicken-cholera scheme to be introduced into New South Wales. MacLaurin had subsequently insisted that he be appointed to this royal commission in his capacity as Government Medical Adviser. Rather than stir up controversy in the press even before this royal commission got under way, Sir Henry Parkes had no choice but to agree.

Beside MacLaurin sat another New South Wales delegate, and yet another man Parkes would rather not have seen on the Commission – Dr William O'Brien Camac Wilkinson, MP. This was the same Dr Camac Wilkinson who had written to the Lands Minister in February decrying Pasteur, decrying anthrax vaccination, and recommending that chicken-cholera be banned from the colony until it was proven safe to do otherwise. It seems that in lobbying the Premier for a seat at the Commission table, Wilkinson had pointed out to Parkes that in his letter of 9 February he had called for 'a full and fair trial' of Pasteur's method and, as a Free-trade supporting member of the Legislative Assembly, he would be the ideal man to look after the government's interests during the Commission's deliberations. Dr Wilkinson would prove to be a pivotal figure in the intriguing events that would soon unfold in relation to Louis Pasteur's bid to win the $10 million rabbit prize.

Wilkinson's letter of 9 February conjures a picture of a pedantic old man but he was in fact just thirty years of age, wiry, fit and energetic. Born in Sydney, the son of a judge, Wilkinson had been awarded his arts degree at the University of Sydney before taking his doctorate in medicine at the University of London in 1884. After graduating, Wilkinson had undertaken postgraduate study in Europe before returning home in 1885 and joining Sydney University's Pathology Department. In October 1885, within months of his return to Sydney, Wilkinson had been elected to the Legislative Assembly as the member for Glebe.

Next to Wilkinson sat the third New South Wales Rabbit Commission delegate. Edward Quin was the proprietor of Tarella Station, located near Wilcannia in western New South Wales, the heart of the most heavily rabbit-infested area in the colony. Quin was here to represent the views and interests of the New South Wales man on the land, the direct victim of the rabbit plague. Throughout the Commission's deliberations, Quin would show himself to be a man obsessed with saving money, especially the money of squatters like himself.

Victoria's delegation was, like that of New South Wales, made up of medical men and a pastoralist. The youngest but most senior member of the Victorian delegation was Professor Harry Brookes Allen. Although just thirty-four years of age, Allen was Dean of Medicine at Melbourne University, where he had trained and qualified, and pathologist with Melbourne Hospital. He was a colourless, businesslike young man with no patience for those he considered his mental inferiors. His fellow Victorian delegate, Alfred Pearson, had been born and educated in Britain before serving as a lecturer in biology at Kutch University in India. Pearson had migrated to Melbourne in 1885 and was now Victoria's Government Chemist. Victoria's third delegate, Edward Lascelles, one of his colony's wealthiest squatters, had been delayed and was expected to arrive in Sydney later in the week.

New Zealand's delegate, Alfred Dillon Bell, had also been delayed and would be in Sydney by the coming weekend. South Australia's delegates had both arrived in good time. Its chief delegate, Dr Edward Stirling, a surgeon, had been Director of the Museum of South Australia since 1884. His colleague, Dr Alexander Paterson, was Government Medical Officer for South Australia.

Queensland's delegate was Dr Joseph Bancroft, President of the Queensland Medical Board and a keen amateur naturalist. Dr Bancroft told his fellow commissioners shortly after this first sitting began: 'Fortunately, we have no rabbits in Queensland at the present time, but every effort will, I gather, be made to help the other colonies to get rid of the plague.'[137] The bunny hordes in New South Wales were in fact moving steadily north and their advance scouts had already crossed the border into southern Queensland, but the Queensland Government did not imagine that they would last long in the sub-tropical Queensland sun.

But it was not rabbits that had prompted the Queensland Government's participation in the Rabbit Commission. While the southern colonies ran some eighty million sheep between them, Queensland was cattle country, the Texas of the southern hemisphere. And the colony was suffering from various cattle diseases, in particular pleuro-pneumonia, for which there was as yet no treatment or preventative vaccine. In March, the knowledgeable and far-sighted Premier of Queensland, Samuel Griffith, who had no qualms about vaccination, had written to his Superintendent of Stock, P.R. Gordon: 'It would be advisable to retain for a term one of Monsieur Pasteur's assistants now on their way to Australia for the purpose of experimenting on the artificial cultivation of the contagion of pleuro-pneumonia.'[138]

Dr Bancroft had subsequently come to Sydney on a secret mission for his colony – to pirate Pasteur's science away from the other colonies when the opportunity presented itself, and to put it to work for Queensland's cattlemen. In the meantime, Bancroft would play the role of a diligent rabbit commissioner.

Tasmania had sent its Chief Inspector of Stock to the Commission. Tasmanian-born, the son of a Royal Navy officer, fifty-one-year-old Thomas Tabart had joined Tasmania's Stock Department in 1870, rising to head the department by 1885. Of all the delegates to the Rabbit Commission, Tabart was the one with the most experience dealing with the rabbit plague on the ground. Tabart's brief from his Premier, Philip Fysh, was to find a solution to Tasmania's rabbit problem if he could, but not at the expense of its 1.5 million fine-wool merino sheep – if in doubt, he was to ensure that Pasteur's chicken-cholera was kept out of the colony.

The tenth man at the table was the Commission's Secretary, Hugh Mahon, a thirty-one-year-old Irishman with a background in journalism. Part of Mahon's job was to use his shorthand-writing skills to note down every word uttered at Commission meetings and hearings. Mahon had spent much of his youth in America. He'd also spent time in an Irish jail with nationalist leader Charles Stewart Parnell for political activism before fleeing to Australia under an assumed name. Mahon had edited a newspaper in Goulburn and been a political reporter for the Sydney *Daily Telegraph* before opening his own paper at Gosford. That publication had gone out of business just as the Government was looking for a Secretary to the Rabbit Commission – at the handsome annual salary of £500 ($200,000 in today's money).

From the outset, the Commission determined not to permit members of the public or the press to attend its sittings. Instead, at the end of each day, Secretary Mahon would brief newsmen who gathered outside, giving them a rundown on what had transpired during the latest sitting. Mahon, himself a journalist, knew the sort of contentious information to keep from reporters – differences of opinion between Commission members, for example. And there would be plenty of those.

It was normal procedure for the government to appoint the presidents of its royal commissions in advance, but Sir Henry Parkes had not wanted to appoint either MacLaurin or Wilkinson to the presidency of this particular commission because of the negative views both had already expressed about Pasteur's chicken-cholera microbes. And the remaining New South Wales delegate, Quin, was far from qualified for the post. Consequently, Parkes left it up to the Rabbit Commission to elect their own president, apparently hoping that they would appoint someone other than MacLaurin or Wilkinson.

Once the ten men had taken their seats at the table, self-important Dr MacLaurin, considering himself the most senior delegate, spoke first, suggesting that the commissioners begin by electing a president. Young Professor Harry Allen from Melbourne quickly nominated MacLaurin; it is likely the pair had agreed on this prior to the Commission sitting. South Australia's Dr Paterson seconded the motion, and Dr MacLaurin was unanimously elected to the presidency. Premier Parkes would not have been pleased.

Harry Allen now asked permission to address the meeting. The tall, bearded young professor from Melbourne University had an aloof manner and a gruff delivery. A doctor of medicine he may have been but he was no scientist. He didn't carry out research, and he didn't encourage his students to do so either. It was his skills as an administrator that had recommended him to the Victorian Government. So much so that, months before, it had also appointed him to the presidency of another royal commission, on the state of Melbourne's sanitation, which was now in progress.

When first invited to sit on the Rabbit Commission, Allen had believed that all the commission was going to do was address Louis Pasteur's proposed method for rabbit destruction, and because of his other commitments he had planned to merely submit a written appraisal of the Pasteur scheme. Once he learned that several other contest entries by colonial applicants were also to be considered, he had changed his mind, asking his government to make arrangements for his university work to be covered by others so as to allow him to take part in the Rabbit Commission.[139]

With MacLaurin's permission, Professor Allen now launched into a long and detailed lecture for the benefit of his fellow Commission members, as if they were first-year medical students. Having read a number of pertinent texts in advance, he had come to this first sitting of the Commission with a broad knowledge of the issues, and with a fixed opinion regarding the advisability of introducing disease to combat the rabbit plague, both of which he proceeded to impart to his colleagues with dreary precision.

Allen wasted little time in getting to the Pasteur submission. Priding himself on a prodigious memory, he had failed to bring any documents to the sitting, including copies of Pasteur's letters to the New South Wales Government – he had already dismissed their content as faulty and inadequate. Monsieur Pasteur's proposals, Allen now told his fellow commissioners, had to be viewed with regard to both their effectiveness and their safety. He acknowledged that chicken-cholera killed rabbits. 'That has been established for many years, and has been demonstrated recently on a large scale by Monsieur Loir in the destruction of rabbits on the Pommery estate. But this fact should not in the least degree justify us propagating disease in these colonies, for the extent of action of such a disease, direct and indirect, is beyond our power of prediction.'[140]

In this one statement, Professor Allen set forth his position regarding Pasteur's proposed method of rabbit eradication. Even though he knew that Pasteur's representatives were at that moment on Australia's shores, ready to conduct experiments to prove to the Commission that chicken-cholera was both effective and safe, Allen wished to ban the Pasteur method, basing his opinion on supposition not on evidence, just as Dr MacLaurin had done in his Health Board reports in February.

There had been much discussion in the colonial press about the Pasteur method ever since it had become known that Pasteur was advocating the use of chicken-cholera against the colonies' rabbits. Predictably, the fact that the disease's name included the word 'cholera' had raised almost hysterical objections in some parts of the community, with the man-in-the-street fearful that Monsieur Pasteur's chicken-cholera would spread to humans. And everyone knew how incredibly deadly cholera epidemics among humans could be.

Opponents of Pasteur's method, the likes of Allen and MacLaurin, had done nothing to dampen those fears. In March, in expectation of the imminent arrival of the Pasteur Mission in Australia, the *Illustrated Sydney News* had run a long article about Pasteur and chicken-cholera based on a book by French authors M.M.V. Babes and A.V. Cornil called *The Microbes*. The author of this newspaper article was unnamed but the language used was suspiciously similar to that employed by Camac Wilkinson in his 9 February letter to the Lands Minister. The article summarised the argument against the Pasteur method by saying, 'The question ought to be gravely considered before a new disease is introduced.' It concluded: 'These dangers by no means should prevent the thorough consideration' of the Pasteur method, 'but point to the necessity of the greatest caution in the conduct of the experiments'.[141]

Allen was riding on the back of this sort of sentiment even though, as a medical man and as an avid reader of medical texts, he knew full well that the only thing that chicken-cholera had in common with cholera was the name, and that it was not in the least harmful to mankind. He also must have known that at this point in time there was no convincing evidence that chicken-cholera spread to birds other than chickens or to animals other than rabbits.

The professor proceeded to then discuss the Pasteur method in detail. He criticised Adrien Loir's Pommery estate experiment as a lost opportunity. Allen told his colleagues that, instead of contaminating all the food of all the Pommery rabbits, Loir should have only infected a few rabbits and then sent them into the burrows to see whether they would pass on the infection to other rabbits. This was in fact exactly what Loir had done; he had only poisoned three bundles of lucerne not all of them – Allen's memory apparently let him down on this score.

Allen's criticism also ignored the fact, pointed out in Pasteur's 5 January letter, that Loir's laboratory experiments in December had demonstrated that uninfected rabbits caught the disease from the carcasses of rabbits that had died from chicken-cholera, and that they too subsequently died. As a result of those experiments there was no question in Pasteur's mind that the disease was highly contagious among rabbits. Allen chose to ignore those results from the December laboratory experiments.

The Melbourne professor went on to reinforce the doubts he'd raised about the contagious nature of the disease by suggesting that it would soon lose its virulence in hot Australian conditions and would not spread beyond those rabbits that had been initially artificially infected. This, he said, would require frequent reapplication of the disease, which defeated its whole purpose and made it no more effective or economical than existing chemical poisons which, he said, were basically the same as chicken-cholera anyway.

Pasteur had already taken into consideration this matter of local conditions and that was why, as Allen would find out the following week, Pasteur was recommending that his team be allowed to conduct experiments in New South Wales on a large inland area to confirm the effectiveness of chicken-cholera in Australian conditions.

Allen went on to spend much time discussing whether it was safe to use chicken-cholera in the colonies to eradicate rabbits. Again relying on memory, he quoted extensively from papers and books by German authors including a Danzig veterinary professor named Zurn; Doctors Huppe, Hitte and Schottelius; 'a distinguished German observer' named Boellinger; as well as Wilhelm Schurz and Robert Koch pupil, Professor Friedrich Loeffler.[142] Allen used the work of the Germans to attack Louis Pasteur's research across the board, even though much of this German material was ten years old and predated Pasteur's most recent discoveries, and came from some of Pasteur's most bitter critics and rivals.

Apart from Pasteur himself, Allen quoted from just one French researcher. This was Joseph-Henri Toussaint, the young Toulouse veterinary professor whose failed anthrax research had opened the door to Pasteur's breakthrough with the anthrax vaccine. Robert Koch still consistently credited Toussaint with discovering the secret of the anthrax vaccine and, following Koch's example, Allen, citing Toussaint as the authority on anthrax vaccine, quoted well out-of-date Toussaint papers from 1880 and 1881 in which he had cast doubt on Pasteur's identification of the chicken-cholera bacillus. Toussaint, 'an observer to whose ability Pasteur has borne testimony', Allen told his fellow commissioners, had established that Pasteur had confused chicken-cholera with acute septicaemia.[143]

Next, Allen turned to Pasteur's more recent discovery – in collaboration with the late Louis Thuillier and assisted by Loir – of the swine fever bacillus. Pasteur, said Allen, had been wrong again. 'Loeffler, of Berlin, states that from his own observations swine plague is due to a minute bacillus, astonishingly like the bacillus of Koch's mouse septicaemia,' said Allen. 'The general conclusion of Dr Loeffler is, that under the name of swine plague several diseases etiologically distinct though perhaps identical in symptoms, have been confused together.'[144]

In fact, as a result of work done in 1884 by Pasteur and Adrien Loir, Pasteur had come to realise that there were indeed several different diseases in pigs that had almost identical symptoms but whose bacilli looked different under the microscope. The great difficulty in diagnosing the disease made the use of a vaccine questionable, and after 1884 Pasteur had dropped his swine fever research and moved into other areas without publishing anything further on the subject. But his original writings, defining swine fever as a solitary bacillus with a figure-of-eight shape, had been published, and so stood exposed and inviting attack by the likes of Loeffler.

In quoting Loeffler, Allen had saved his big gun for last. For, as the medical men in the room knew, Loeffler, together with Wilhelm Schurz, had in 1882 isolated the organism responsible for glanders, a disease in horses. More famously, two years later, in the year he became a pupil of Koch in Berlin, Loeffler had combined with Edwin Klebs to find the cause of diphtheria. In theory then, Loeffler should have known what he was talking about.

Why had Professor Allen apparently come to the Rabbit Commission determined to destroy the credibility of both Pasteur and his rabbit competition entry? Perhaps he did not want to see £25,000 of Australian money going to a Frenchman. There were still people alive in Britain's Australian colonies who could remember the joyous day in 1816 when the news arrived from England that

Napoleon Bonaparte had been defeated at the Battle of Waterloo on 18 June 1815.

But was this Australian-born professor's bias against and resistance to Pasteur and his methods really borne out of anti-French sentiment? Pasteur and other Frenchmen came to believe so. Yet the fine detail of Allen's destructive address to the Commission gives the best clue to his motive. All his references but one came from German figures, and the one exception was a Frenchman opposed to Pasteur's views and given some credit by Robert Koch and his followers. Allen was clearly a supporter of the German view. But not necessarily because he favoured Germany and opposed France or the French. Over the past few years, sponsored by Robert Koch and his disciples, there had been a growing movement against inoculation. This anti-vaccinist movement had gained many followers in many parts of the world, but most particularly in Germany and Britain.

Edward Jenner's advocacy of inoculation had seen adoption of the process across Europe for vaccination against smallpox, with Prussia, ironically, pioneering it in 1835. It was 1853 before legislation was introduced to make smallpox inoculation compulsory for all children in England and Wales, with parents fined £1 if they failed to comply. South Australia, an innovator in so many areas, was the first Australian colony to make inoculation for smallpox compulsory for children, in 1872. Victoria in 1874, Western Australia in 1878, and Tasmania in 1882 followed its example.

New South Wales had toyed with smallpox vaccination as far back as 1804 when surgeons of the New South Wales Corps had inoculated children in the fledgling colony. The practice had fallen into abeyance until voluntary smallpox inoculation was introduced by the New South Wales Government in 1861. Twenty years later it offered doctors the necessary materials at government expense, without making smallpox vaccination compulsory.[145] In 1881, the first year of this free smallpox inoculation program in New South Wales, almost every child was vaccinated. But by 1887, of 37,236 children born in the colony that year, only 3045 were vaccinated.[146] Numbers would drop to 2095 in 1888. While the Medical Adviser to the New South Wales Government put this down to 'increasing apathy and carelessness' on the part of parents,[147] the real reason for this significant decline in inoculations lay with the doctors of the colony, many of whom had developed grave doubts about the safety of the inoculation process.

Those doubts had begun in Prussia and spread across Europe after a particularly virulent smallpox epidemic had swept across the Continent in 1871–72. It was news of this same epidemic that had prompted the South Australian parliament to make smallpox inoculation compulsory in 1872. In Prussia, the outbreak had the opposite affect. There, where smallpox inoculation had been practised for thirty-seven years, the 1871–72 epidemic killed 69,000 people, many of them young children – all, it would be claimed, had been vaccinated or re-vaccinated against smallpox. This caused Prussian medical men to turn against inoculation, believing that it had either been ineffective or had contributed to the epidemic's death toll. Robert Koch was one of them; he had rushed back to his private practice at Rakwitz in 1871 following the war with France to help the locals cope with the outbreak, and then worked as a district medical officer in Wollstein from the following year and dealt with the tragic aftermath of the epidemic.

An anti-vaccinist movement had subsequently sprung up, opposing vaccination on the grounds that it was unproven and unsafe, with many anti-vaccinists making claims against vaccination that were both dubious and hysterical, as they tried to destroy 'the hotly-disputed Jennerian dogma'.[148] This negative attitude to vaccination and vaccines was pervading the medical community of the Australasian colonies at the very time that Pasteur sent his team to Australia. It is clear that Professor Harry Allen was either anti-vaccination or pro-German bacteriology or both, which made him the committed adversary of Pasteur.

Not that the non-medical members of the Rabbit Commission had much of an idea about these two warring worlds of the vaccinists, led by Pasteur, and the anti-vaccinists, led by Koch. Nor did they realise that Allen was promoting a one-sided view, the Koch view, of Pasteur's science. In fact, to one of the commissioners, Wilcannia farmer Edward Quin, Allen's address was positively illuminating.

He moved that it be printed, for the edification of both the current members of the Commission and those delegates who were still on their way to Sydney, and his colleagues voted in agreement.

Despite his undisguised antagonism toward Pasteur and his recommendations, Professor Allen, being an intelligent man, knew that there was a political will at the highest level of the New South Wales Government for the Pasteur method of rabbit eradication to be tested. So before he gave up the floor, he added, 'If it be determined to conduct a series of experiments, the operations may be divided into three groups – laboratory experiments, experiments on an island preferably already infested with rabbits, and experiments on a portion of the mainland under the natural conditions which will prevail in the ordinary cases.' To sound evenhanded, Allen was to concede that Monsieur Pasteur's representatives might have valuable evidence to submit with which he was not familiar and also observed, giving himself a safety net: 'The whole domain of bacteriology is still beset with obscurity; facts are often doubtful, and inferences uncertain.'[149]

Mr Quin now begged to differ from Professor Allen on one point. He felt that experiments on an island were not a good idea as the conditions of climate and vegetation on islands and even on the mainland near the coast were quite different from those inland where the rabbit plague was greatest. He therefore proposed that if any experiments were carried out they should take place inland.

Dr Paterson from Adelaide spoke next, presenting a report that he had prepared for the South Australian Government on Professor Watson's rabbit scab disease which, he had concluded, deserved further examination. Despite that opinion, he said, 'I am somewhat sorry to say that I do not have the very sanguine expectations that have been formed by the promoters of the various schemes – Monsieur Pasteur's scheme, the Tintinallogy scheme (Ellis–Butcher), or the scheme of Professor Watson.'[150] Far from coming to the Commission with an open mind, Paterson did not believe that any of the schemes vying for the rabbit prize would work, and had no expectations of seeing or hearing anything that would change his mind.

Victoria's Government Chemist Alfred Pearson told his fellow commissioners that he didn't believe that Pasteur's disease was the best method with which to begin experiments. Chicken-cholera killed rabbits far too quickly for his liking. As far as he was concerned, a rabbit had to be alive to communicate a disease to another rabbit. Like Professor Allen, Pearson had failed to note Pasteur's observations about Loir's experiments in the laboratory and in the field which had shown that uninfected rabbits had caught chicken-cholera from the carcasses of rabbits that had died from the disease, and these rabbits had also died.

Then again, while Allen was a voracious reader perhaps Pearson was not, and had not read in detail all of the material sent by Pasteur. Or perhaps, like Allen, Pearson did not want to acknowledge this information. For as time was to prove, Pearson had an ulterior motive for wanting all the biological methods to be dismissed by the Commission so that another method of rabbit eradication could be employed. 'Before we try Pasteur's method,' he told his colleagues at the Commission table, 'we should endeavour to get something better in the colonies, if such a thing exists.'[151] Only much later would Pearson reveal what he considered to be that 'something better' – it was an alternative in which he had a vested interest.

Dr Bancroft, the secret agent of the Queensland cattle industry, had clearly not read all the Pasteur material either, for he accepted Professor Allen's line on contagion of chicken-cholera among rabbits. 'It appears this disease of Monsieur Pasteur's propagates itself among fowls but not amongst rabbits,' Bancroft commented. He advocated taking time over experiments – to keep the Pasteur team in Australia for as long as possible and so suit Queensland's undisclosed agenda. And he supported the idea of conducting the chicken-cholera experiments on an island.[152]

Tasmania's commissioner, Alfred Tabart, established his position from the moment the President gave him the opportunity to speak. Tabart declared that he considered the rabbit pest 'the great question of the Australasian colonies', and one that had to be dealt with vigorously and stringently. 'At the same time, I do not consider it judicious to allow any foreign disease to be introduced, with the view of rabbit extermination, until it has been conclusively proved beyond a shadow of a doubt that the disease introduced will not be communicated to sheep or other animals.' Like Dr Bancroft, Tabart also supported Professor Allen's proposal that any experimentation with imported diseases be carried out on an island.[153]

Through the morning and during the sitting that followed lunch, Dr Camac Wilkinson MP had been quietly biding his time, allowing the other members to have their say and so expose their prejudices and their ignorance without interruption or comment from him. Finally, late in the afternoon, he took his turn. Dr Wilkinson began by complimenting Professor Allen 'on the delivery of an able and exhaustive address' that morning. But, he said, he took exception to some of the points that Allen had made.[154]

Allen, Wilkinson said, had likened chicken-cholera to the action of phosphorous or arsenic. It was nothing like it, he said. Allen, Wilkinson went on, had dismissed the Pommery experiment because it had involved the communication of chicken-cholera through food and the disease had not apparently spread from rabbit to rabbit. If, therefore, the disease was not contagious, said Wilkinson perversely, this could be an advantage because there would be no danger of it spreading to animals other than rabbits. Wilkinson then rather pompously informed his fellow commissioners that he had personal experience of chicken-cholera: 'I know that this disease in fowls kills mice. I had an opportunity of making experiments in Berlin under Koch, and I remember perfectly having introduced these germs into mice and killing them. The recommendations that Professor Allen has made are practically the same as those which I made in a letter addressed to the Government some weeks ago.'[155]

This announcement, that he had studied under the famous German bacteriologist Robert Koch, was designed to impress his colleagues. Wilkinson's claim was also intended to knock Harry Brookes Allen from the lofty pedestal of his own creation – for, as Wilkinson knew, Allen had never studied outside Australia. But had Camac Wilkinson really been a pupil of Koch? Wilkinson had completed his medical studies in Europe in 1884–85 under German tutors and he liked to give the impression that he had been a pupil of the famous Robert Koch during that time. The records show that Wilkinson had attended Vienna University and Strasbourg University during this period.[156] He undertook no formal training in Berlin.

Perhaps Wilkinson did make a brief visit to the German capital during 1884–85, and, with an introduction from Vienna, was able to conduct experiments in Koch's Berlin laboratory. But Camac Wilkinson was a man prone to exaggeration, and to glossing the truth – stock in trade for a politician – possessing the necessary skills of self-promotion to win election to parliament at the age of twenty-seven and to be re-elected two years later. Wilkinson's carelessness with the truth had begun when he went to England to study for his doctorate. There, he had given a false date of birth, making himself seem a little older than he actually was, and gave his birth-place as well-to-do Glebe Point, where his parents moved to in later years, when in fact he'd been born in working-class Enfield.[157] These are minor lies, it is true, but they indicate that Wilkinson had a habit of trying to improve on fact when it came to his background.

Wilkinson was clever enough to tell intercontinental fibs – few people then in Australia had grounds to question his assertion that he had been a pupil of Koch in Germany. Certainly, none of the members of the Rabbit Commission was in a position to do so – unlike Adrien Loir, who was to say that Wilkinson '*claimed* to have studied under Koch'.[158] Clearly, Loir had his doubts about this claim. Wilkinson's track record suggests that, while he was without doubt a disciple of Koch and his bacteriological teachings, he may well have invented the story of studying under him.

Having staked a claim as the most knowledgeable man in the room when it came to chicken-cholera, Wilkinson went on to demolish Professor Allen's lecture of that morning. Allen, said Wilkinson, had referred to numerous (German) experiments conducted ten years before in 1878. But, he said, 'The whole question of bacteria was then in its infancy, and when experiments were made no precautions were made to prevent contamination.' In the same vein, there was Allen's use of Toussaint's flawed and outdated work to attack Pasteur, which Wilkinson proceeded to accurately characterise:

'Professor Allen's account of Toussaint's experiments, and experiments of a similar kind, may have a historical interest but cannot be of any practical value to the Commission in its important work.'[159]

But if his listeners thought that Wilkinson was taking Pasteur's side, he soon dispelled this impression. Wilkinson's focus was on destroying Allen's credibility but he was no supporter of Pasteur. 'The Commission can place no trust in such statements,' he said of Allen's claim that the symptoms of swine plague and chicken-cholera were so similar that it was almost impossible for bacteriologists to differentiate between the diseases. 'If this were true,' said Wilkinson, 'it would mean all the work of Koch and other leading bacteriologists was to be upset. Klein has proved, beyond all question, that Pasteur was wrong, that his conclusions were drawn from experiments open to the gravest objections.'[160] Pasteur, like Professor Allen, was in Wilkinson's sights. All Wilkinson's ammunition had been manufactured by the Koch bacteriological school. Faulty ammunition at that.

Squatter Edward Quin moved that Wilkinson's address also be printed, to which Queensland's Dr Bancroft, who seems to have been more impressed by Professor Allen's presentation than by Wilkinson's egotistical rebuttal of it, responded, 'I shall be glad to second that. Facts, such as those mentioned by Professor Allen, disappear quickly from people's minds.' The motion was carried.[161] Quin next moved that the Commission recommend to the New South Wales Government that it provide both an island and an inland tract of land where experiments with rabbit diseases could be safely carried out, and that the government provide the necessary funds to cover experimentation. This too was agreed unanimously.

Throughout the day, members of the Commission had criticised Pasteur, his science, and the method of rabbit eradication that he was proposing for New South Wales, without Pasteur or his representatives being given the opportunity to offer a defence, and before they could even demonstrate their method. Only with the sitting about to conclude did South Australia's Dr Stirling move that Pasteur's representatives be invited to meet the Commission at 10.00 am the following day, when they could at last speak for themselves. The motion was agreed.

At 5.30 pm, Commission President Dr MacLaurin, whose only contribution throughout the day had been to chair the meeting, adjourned the Commission until the following day, when Adrien Loir would at last be able to put his uncle's case.

11

FIFTEEN HUNDRED ENTRIES

AT 10.00 AM ON Tuesday 17 April, the Rabbit Commission met for the second time. As Dr MacLaurin failed to put in an appearance, young Professor Harry Allen took the chair as Acting President with the agreement of the majority of his fellow commissioners.

From the attitudes displayed by some of the commissioners on the first sitting day, it seems they believed that their primary role was to prevent the employment of the Pasteur scheme in Australia. All the focus on Pasteur's method seemed to ignore the fact that there were other entries in the international competition. In fact, there had been more than 1500 of them. Frank Abigail's call to the world had brought forth submissions from around the globe.

Fourteen hundred of those entries turned out to be for a mind-boggling array of rabbit trapping and killing devices – some ingenious, some unworkable, and all of them outside the purview of the Commission, which was required by the terms of reference originally laid down by Frank Abigail to only consider biological methods of rabbit destruction. This left 115 biological schemes – forty-two from the Australian colonies and six from New Zealand, twenty from Britain and Ireland, seventeen from the United States, twelve from France, and just two from Germany, with the remainder coming from most of the other countries of Europe and one each from Fiji and Netherlands' India.

The Commission was to agree that the Ellis–Butcher scheme and that of Professor Watson's, both of which Lands Minister Garrett had felt showed promise, should be further inquired into, along with the Pasteur scheme, and experiments conducted accordingly. Commission Secretary Mahon would also be instructed by the commissioners to write to several other entrants whose brief submissions were interesting enough to explore if more details were provided.

One of those latter entries had come from James P. West and Henry J.W. Raphael of Oxford Street, London. West and Raphael, who dabbled in inventive enterprises, had proposed the use of an existing rabbit disease known as nasal catarrh. They claimed to have personally cultivated microbes of this disease and using these, to have conducted experiments on rabbits that had proven its fatal effects. They even claimed that rabbits that survived an initial nasal catarrh infection did not develop immunity and would succumb to a fresh outbreak of the disease. West and Raphael had said they would prefer to keep their method of application of this disease to themselves until the Commission invited them to take their entry to the next stage. The Commission was sufficiently interested to send a telegram to Agent-General Sir Daniel Cooper in London, asking for him to arrange to have West and Raphael forward microbe cultures of their disease to the Commission for experimentation.[162]

Secretary Mahon would write to several other contestants asking for more details of their proposed schemes. One letter was sent to American entrant J.H. Richardson of Colorado. Richardson, a naturalist, had described a severe disease that had recently devastated the rabbit population of Colorado, but had given no comprehensive details. Letters were also written to an entrant from Sydney and another from Cairns in Queensland whose submissions were interesting enough to warrant further details; and similarly to two more French entrants: Monsieur M.A. Laplanche of Fismes, and Monsieur M. Lanferon of Nevers.

In addition, at the suggestion of either Dr Wilkinson or Professor Allen or both, who would have been disappointed by the poor

German response to the call for rabbit eradication entries, Secretary Mahon was instructed to write to Dr Hugo Ribbert, the thirty-five-year-old professor of pathology at Berlin University, and to Dr Friedrich Loeffler, who'd come out from under Robert Koch's wing to take up the post of professor of hygiene at Greifswald University. Both Ribbert and Loeffler were asked to provide further details of rabbit diseases they'd described in recently published scientific papers, in particular rabbit septicaemia. It would take many months for all the correspondence to be dealt with. In the meantime, the Commission agreed that entries offering immediate demonstration should be explored in depth, starting with the Pasteur method.

As the Rabbit Commission convened in the Executive Council Chamber on the morning of 17 April, Adrien Loir's English associate Dr Frank Hinds was waiting in the corridor on the third level of the Colonial Secretary's Department. For the past two weeks Hinds had been living at Sydney's Union Club, calling at the Colonial Secretary's office every day, and earlier that morning he'd been advised by the office staff that the Rabbit Commission was in session and wished to see Monsieur Pasteur's representatives at 10.00 o'clock.

Before the commissioners asked for Hinds to be ushered in, they worked through an amount of business. They agreed that Pasteur's scheme and any other diseases not then in existence in Australia should be tested on an island without delay, and that experiments on a site in the interior of New South Wales would be held off until their safety and practical value had been demonstrated. It was agreed that experiments with any disease already existing in Australia could begin in the interior at once. At Mr Quin's suggestion, it was agreed that the government would be asked to immediately fence off with rabbit-proof fencing, at public expense, an area not exceeding 10,000 acres (4000 hectares) in an area in the interior of the colony currently infested with rabbits, as the location for approved experiments.

It was also agreed that a recommendation be passed on at once to the government that Monsieur Pasteur's representatives be allowed to introduce chicken-cholera microbes into New South Wales, provided they did not undertake any experiments with those microbes without the sanction of the Commission. The government would also be asked to immediately provide a laboratory where experiments involving the Pasteur scheme and similar schemes could be carried out.

Frank Hinds was then called in. Hinds' appearance before the eight commissioners was brief. When asked where Monsieur Loir and his French associate were, Hinds replied that they were in Melbourne awaiting permission to bring Monsieur Pasteur's chicken-cholera microbes into New South Wales so they could conduct experiments before the Commission. When informed that the microbes could now be imported into the colony, Hinds said that that being the case, Loir and Germont would probably arrive in Sydney on Thursday. So Professor Allen asked Hinds to bring the Frenchmen before the Commission on the following Monday at 10.00 for interview. Hinds hurried away to cable Adrien Loir in Melbourne and tell him to come at once.

12

LOST IN TRANSLATION

THIRTY MINUTES AFTER NOON, the Sydney Express rolled up to the redbrick station buildings of Sydney's Redfern Terminus – which stood very near to where its successor, today's sprawling Central Station, was built in 1906. Sliding in beside the main platform, the big main-line locomotive slowed to a halt with a squeal of metal on metal, bringing the string of carriages behind it to a shuddering stop. As steam hissed from the locomotive, carriage doors flew open and hundreds of weary passengers gratefully stepped out onto the platform. Behind them lay a nineteen-and-a-half-hour journey from Melbourne. From a First Class carriage, immaculately dressed Adrien Loir and bulky François Germont alighted as fellow passengers barged past.

Loir and Germont had dropped everything the moment Frank Hinds' telegram had reached them in Melbourne, and had caught Tuesday night's train. This brought them to Sydney on the Wednesday, a day earlier than Hinds had initially anticipated. As Loir was leaving Melbourne, he had sent a telegram to Noumea asking for the box of chicken-cholera microbes in storage there to be sent to him in Sydney. As for the second box containing microbes that he'd proved were still virulent, he had left it in safe-keeping in Melbourne.

Loir did not entirely trust Hinds and until he had satisfied himself that it was safe to bring the proven microbes up to Sydney without the risk of the New South Wales Government confiscating them, they would remain in the Victorian capital.

Frank Hinds was not at the station to meet Loir and Germont. Probably too embarrassed to turn up because of his poor French, he had instead alerted the French Consulate in Sydney of the pair's imminent arrival. The Consul was away on holiday at the time but the Consular Secretary, the Marquis De Rostain, had come to meet Loir and Germont at the train. An elderly gentleman with a grey beard and a monocle, De Rostain kissed his fellow countrymen on the cheek, pumped their hands, and enthusiastically offered to help them in any way he could.

When Loir explained that they were hampered by the lack of a competent interpreter, the Marquis said that he would personally take them to meet the instigator of the rabbit eradication competition, Mines Secretary Abigail, and act as their interpreter. A delighted Loir immediately accepted the offer. Loir and Germont had their baggage loaded onto the Marquis' carriage, and the three of them set off at once for the Mines Department in Macquarie Street.

The streets of this city of 400,000 people were bustling with pedestrians, tramcars and horse-drawn traffic of every description. Flags flew from every building and red, white and blue bunting was draped across frontages and between lamp-posts. Shopfronts were decorated with patriotic displays and in major streets temporary triumphal arches stretched from pavement to pavement. The reason for this gay display was that prosperous Sydney was celebrating the colony's centennial – one hundred years since the arrival of the First Fleet from England with its 1500 convicts, soldiers and settlers; one hundred years since the founding of Sydney. While the centennial mood would last throughout the year, the major celebratory events had taken place back in January, with numerous dinners, balls, laying of foundation stones for new buildings, and excuses for political speeches.

Loir, Germont and the Marquis De Rostain pulled up outside the Mines Department this April day. At last, after spending frustrating weeks in Melbourne, Loir felt he was getting somewhere.

Frank Abigail was in his office when the trio of Frenchmen arrived unannounced. Thrilled to meet the nephew of Louis Pasteur, Abigail immediately invited the party into his room. The young minister looked a little French himself, with his long face, neat moustache and goatee beard, and there was something quite Gallic about the dramatic way he threw his hands around as he spoke. Abigail was soon telling his visitors of his great disappointment at the way the rabbit competition that he'd created had been taken from his hands. As the Marquis attempted to keep pace with his translation for his two young companions, Abigail then warned the Frenchmen that there were a number of men connected with the Rabbit Commission who were of the opinion that Pasteur's method was of no value and did not even deserve to be tested.[163]

Sadly, De Rostain's translation skills were wanting. He wrongly told Loir and Germont that Abigail himself was of the opinion that Pasteur's method was of no value and did not even deserve to be tested. Loir came to his feet, shook the minister's hand, thanked him for his time, and led his companions from the office, leaving Abigail taken aback. Loir and Germont were furious, believing that they'd come all this way only for the initiator of the competition to insult the work of Louis Pasteur to their faces. Loir was to later say that it did not even occur to Germont or himself at the time to doubt the English language skills of the Marquis.[164]

Believing that Abigail was against testing the Pasteur method when in fact he was all for it, the two dejected Frenchmen rode with De Rostain to the Union Club in Phillip Street where Frank Hinds was waiting for them. Hinds was also crestfallen when the Marquis told him that Minister Abigail had dismissed the Pasteur competition entry out of hand. Hinds, who had no dedication to the Pasteur Mission having only come along for the ride, immediately decided to apply for the vacant post of hospital director with one of Sydney's hospitals. If his bid for the hospital job failed, he said, he would go home to England.

Loir was, in his own words, 'completely discouraged'.[165] He did not want to let his uncle down but what was he to do? For the moment he sought and received one month's honorary membership of the Union Club, which gave him somewhere to stay in the short term. He had money enough now, courtesy of the Victoria Brewery, to fund a protracted campaign in New South Wales so he never seriously thought of going home, despite the apparent rebuff by Minister Abigail.

Only later would Loir and Germont realise that the Marquis had set them off on the wrong foot with his mistranslation. Even though this misunderstanding had yet to be clarified for them, Loir and Germont knew that their lack of English was an impediment that they must overcome, and quickly. So the pair agreed to split up. Loir would remain at the Union Club while Germont found accommodation elsewhere. They made a pact: from now on neither of them would utter a word in French; instead they would speak English exclusively and force themselves to learn the local language. One way or another, they were determined not to leave Australia until they had succeeded in accomplishing their mission for Louis Pasteur. 'In spite of everything,' Loir was to say, 'Germont and I were still hopeful.'[166]

Frank Hinds informed Loir and Germont that the Rabbit Commission wished to see them the following Monday. This meeting, Loir assumed, would only be a formality, during which the Commission would confirm what De Rostain had led them to believe constituted the attitude of Minister Abigail and the New South Wales Government – that the Pasteur method would not be accepted for testing in the colony. Still, it was contact with the Commission and, hoping to perhaps convince the commissioners to allow them to demonstrate the Pasteur method after all, Loir told his companions that all three of them would keep the appointment.

At 10.00 on the morning of Monday 23 April, Loir, Germont and Hinds were waiting in the gloomy, stone-tiled corridor at the Colonial Secretary's Department as the members of the Commission sat down to meet just along the corridor.

For the first time, every member of the Commission was present at this 23 April sitting. With dour Scotsman Dr MacLaurin once more taking the President's chair, two additional commissioners joined their colleagues at the table.

Forty-year-old Edward Lascelles, Victoria's third commissioner, had been born in Tasmania and as a youth had gone into the agricultural business with an uncle at Geelong in southwestern Victoria. By the time he was in his thirties, Lascelles was a rich man. Lascelles now controlled half a million acres (125,000 hectares) of rural land including large runs in the Mallee district which were currently greatly troubled by rabbits. A born entrepreneur, Lascelles had diversified. Among other ventures, he imported and manufactured wire mesh fencing. He was to tell the Commission, 'I think I have erected more netting fences than any other man in the colonies.'[167] And, together with his neighbour Andrew Anderson, he manufactured the Lascelles and Anderson Patent Rabbit Poison Distributor. This horse-drawn contraption featured a scarifier that cut furrows in the ground and a hopper on top from which a pipe fed wheat poisoned with phosphorous into the furrow. Rabbits were attracted to the freshly turned soil, ate the poisoned wheat distributed by the Lascelles and Anderson machine, and died.

Life was rosy for Edward Lascelles right now – he had only been married a matter of months and business was good despite too many rabbits and not enough rain. It was so good that he was planning a major rural subdivision which, within two years, would see him replace sheep with wheat in the Mallee; he would even create his own town to service the wheat farmers who bought land from him – Hopetoun.

New Zealand's delegate to the Rabbit Commission, Alfred 'Fred' Dillon Bell, had also arrived in Sydney, and he too took his seat at the Commission table for the first time this Monday morning. At first glance, Fred Bell's qualifications to be New Zealand's sole delegate to the Rabbit Commission may have seemed inadequate. He ran a number of farms in New Zealand for his father, living on one of the South Island properties, Shag Valley Station in eastern Otago. In sending Fred Bell to Sydney, New Zealand authorities would have pointed out that those properties were suffering from the rabbit plague. Besides, Fred's connections were impeccable: he was the second son of Sir Francis Dillon Bell, the New Zealand Agent-General to London who had only recently met with Louis Pasteur in Paris. And Fred's elder brother, Francis Dillon Bell Junior, was New Zealand's Crown Solicitor; he would be elected Mayor of Auckland within three years and several decades later would become New Zealand's first locally born Prime Minister.

Fred Dillon Bell was in no hurry to return to New Zealand. He had come to Sydney prepared to stay as long as it took for the Commission to do its business, and determined to learn as much as he could from the scientists who would be involved with the Commission. A man with an insatiable curiosity and a reputation for being more interested in science than in sheep, Bell had rigged up New Zealand's first working telephone between farm buildings on Shag Valley Station. He knew all about the problems that rabbits were causing New Zealand's sheep farmers – the colony had exported sixty million rabbit skins over the past six years. But the skin trade had never fully paid the cost of killing the rabbits that abounded in the green valleys of the North and South Islands. One estimate put the colony close to a million pounds sterling out of pocket by 1888 because of the rabbit pest, even after the sale of skins was taken into account.[168]

Commission President Dr MacLaurin called the latest sitting of the Rabbit Commission to order. But before MacLaurin asked for Monsieur Pasteur's waiting representatives to be brought in, the Commission dealt with several letters from the government. One stated that the Lands Department had received two offers from squatters prepared to make their properties available for research purposes. One property, Roto Station, was near Hillston. The other, Tarella, was outside Wilcannia. The commissioners voted to accept the Tarella offer. Edward Quin abstained from voting because Tarella was his property, and he had quite deliberately offered it to the Lands Minister for use by the Commission. Earlier, it had been Quin who had convinced the Commission to recommend that the government pay for rabbit-proof fencing on 10,000 acres on the property chosen for inland experiments. Thrifty Edward Quin now stood to gain kilometres of free fencing, courtesy of the Rabbit Commission and the public purse. His fellow commissioners were fully aware of this, yet still they voted to use his property and not the other on offer. This was to be the first of many questionable decisions by the Commission.

The commissioners now learned from Dr MacLaurin that the Board of Health had not yet been approached by the Lands Minister – Premier Parkes at the time – for formal approval of the use of the Government Laboratory for chicken-cholera and other experiments. As if to spite the Premier, the Commission promptly decided to approach the Linnean Society for the use of its laboratory at Elizabeth Bay.

It was also agreed that a telegram should be sent to a Dr Oscar Katz, requesting his attendance before the Commission that afternoon. This was the first official mention of Oscar Katz's name; the Commission minutes do not record any formal discussion taking place about Katz prior to this. While his sudden appearance on the scene was unheralded, it is clear that outside the Commission members had agreed among themselves to offer Dr Katz the position of the Commission's Chief Expert Officer, who would oversee all experiments on behalf of the Commission.

Once these items of business had been attended to, the Commission's President summoned Monsieur Pasteur's representatives. Adrien Loir, François Germont and Frank Hinds were escorted into the chamber. For a moment, the trio paused on the threshold, taking in the room. The green walls were lined with paintings depicting British figures who were mostly unknown to Loir and Germont, but the large portrait above the marble fireplace to their left was unmistakably of a young Queen Victoria. There were numerous bronze busts set around the room on marble plinths, with that of Shakespeare the most recognisable and lifelike. A long table directly in front of the Pasteur team, the Executive Council table, was not occupied. The members of the Rabbit Commission were sitting on the far side of a committee table to the visitors' right, beneath a portrait of Captain James Cook.

Pasteur's representatives crossed the thick red carpet and were introduced to all eleven members of the Commission, then took their seats on straight-backed green leather chairs facing them. This was the commissioners' first sight of Pasteur's protégés. Fresh-faced Loir, despite his moustache, was seen as 'a comparatively young man'.[169] Germont was taller than Loir by a number of centimetres and broader; while only a few years senior to the party's leader, he looked considerably older, with an 'eminent'[170] appearance enhanced by moustache and goatee beard.

Loir and Germont in turn studied the faces of these bearded and moustachioed men who would decide the fate of Louis Pasteur's bid to win the $10 million prize, not knowing at that time that every commissioner had come to the Commission with his own agenda. It would only later become apparent that Dr MacLaurin wanted to control everything, and that Professor Allen wanted to steer his colleagues toward his way of thinking. That Dr Wilkinson wanted to dominate and impress. That Doctors Stirling and Paterson wanted to play safe, while Fred Bell wanted to play at being a scientist. That Chief Stock Inspector Tabart wanted to safeguard his colony's sheep. That Dr Bancroft wanted to secure Pasteur's science for the benefit of his colony's cattle. That Alfred Pearson wanted the Commission to adopt his preferred rabbit remedy. That squatter Quin wanted to save money while, conversely, entrepreneur Lascelles wanted to make money.

Pasteur's representatives must have wondered just what qualified these colonial gentlemen to sit in judgment on the Pasteur proposal. Over time, Loir and his colleagues would learn that while some commissioners had been involved in the field of biology during their medical training, apart from Wilkinson's brief laboratory experiments in Germany, none of the commissioners had any experience in biological research. Almost all their relevant knowledge came from what they had read or been told, with most of that information coming from Pasteur's rivals. Had Pasteur's team members then known of the judges' deficiencies, they might well have said that expecting these particular gentlemen to adjudicate in this competition was like asking a team of abattoir slaughtermen to judge the work of a brain surgeon.

From their first meeting, these commissioners had adhered to a single unifying tenet – a determination not to take risks. The rabbits that had given rise to the Commission were themselves proof of the dangers of introducing foreign elements into the Australian ecosystem. Today Australia is free of many diseases that afflict other continents primarily because successive Australian governments have maintained strict quarantine standards. The rabbit commissioners of 1888 could have approached their task with the attitude that they were embarking on a cutting-edge scientific project – if Pasteur's chicken-cholera method of vermin destruction was successfully employed in Australia it would be a world first, and could lead the way to its widespread adoption internationally. But unlike Louis Pasteur, the commissioners were not intrepid pioneers when it came to the advancement of science.

For decades, Pasteur had operated on the principle that without risk there can be no scientific breakthroughs. Little more than a year before this Rabbit Commission sitting, on 11 January 1887, Pasteur's loyal and dedicated associate Dr Jacques-Joseph Grancher had stood before a meeting of the Academy of Medicine in Paris and defended Pasteur against the attacks of Dr Michel Peter. Peter, Adrien Loir's former medical tutor, had condemned Pasteur for imperilling the lives of the people he was injecting with rabies vaccine. In defending Pasteur, Grancher had acknowledged that there were risks involved. But, he had said, there were risks inherent in all medical therapy; moreover, a person takes a risk every time they cross the street. Grancher had argued that without taking risks, science could not advance.

Like Grancher, Adrien Loir had been imbued with his uncle's risk-taking attitude, as evidenced by the way he had volunteered to be a rabies vaccine guinea pig. These divergent attitudes to risk held by the commissioners and the Pasteur representatives were at the core of the wide gulf that was opening between Pasteur and the Rabbit Commission.

Now Commission President MacLaurin invited Secretary Mahon to read aloud a letter sent direct to him by Louis Pasteur on 20 February. As with his 8 January letter to Mines Secretary Abigail, Pasteur had headed his letter, 'The Pasteur Institute, Paris' even though the institute was at that time nothing more than a building site at Grenelle; Pasteur was still operating from 45 Rue d'Ulm.[171]

Pasteur began this latest letter by referring to his communication to *Le Temps* in November. After sending that letter to the newspaper, Pasteur wrote, 'I made various experiments, both in my laboratory and on a more extensive scale, which allowed me to believe that the application of chicken-cholera to the destruction of rabbits would answer all the conditions of the scheme of reward proposed by the Government of New South Wales.' Consequently, said Pasteur, he had decided to send to Sydney, at his own expense, three young scientists entrusted with the task of making all the necessary demonstrations to prove that his scheme for rabbit destruction worked. 'Should the reward be adjudged to me when the Commission arrives at its decision, the delegates shall be bound to reveal to the Commission, in all its details, the process which shall have to be applied, especially certain particulars without which it could not be made practical.'[172]

The medical men among the commissioners would have been quick to note Pasteur's implication: that there was some secret to the application of chicken-cholera as an effective rabbit killer, a secret known only to Pasteur and his representatives.

Pasteur had continued: 'In the scheme of reward of the Government of Sydney, published in Europe at the end of the year 1887, it is stated "that the scheme should not be judged by the Commission nominated for that purpose, until after a year's trial". I do not believe that so long a time is necessary, and I am persuaded that the Commission will share my opinion on this point. I think that the Commission entrusted with adjudging the reward will be able to arrive at a conclusion after a month or six weeks' trial.'[173]

A month or six weeks? Pasteur seemed to be in one heck of a hurry to lay his hands on the £25,000 prize. On hearing this, members of the Commission would have glanced at each other with raised eyebrows. The government proclamation, which had been authorised by the Governor of the colony, Lord Carrington, at a meeting of the colony's Executive Council, had quite specifically provided for a year's trial. These were the contest's rules and the Commission had no power to alter this condition even had its members wanted to do so. This attempt to push the Commission into a hasty decision would not have impressed the commissioners. As it was, most were already dubious about Pasteur's chicken-cholera scheme, if not steadfastly opposed to it.

Only Adrien Loir knew that Pasteur was desperate for the prize so that he could bolster the fund for his new institute. That desperation had shown itself in January when Pasteur had been in such a hurry to send Loir and his team to Australia that, not only had he sent them uninvited, he had overlooked the fact that a French shipping line – the Messagéries Maritimes Line, the same line whose captain had taken Loir's microbe box to Noumea – sailed regularly to Australia and New Caledonia from Marseilles. As Loir was now aware, he need not have lingered in Naples waiting for the Orient Line ship.

Loir knew that Pasteur was being presumptuous in suggesting the Commission award the prize after just four or six weeks of trials. Yet Loir could sympathise with his uncle. Pasteur did not object to commissions appraising his work. In France, and even in Germany, Pasteur had actively encouraged commissions of experts to view his demonstrations of this procedure or that vaccine, always confident that his work would come up trumps when put to the test. Those past commissions – whether they consisted of members of the French Academy of Sciences or the Academy of Medicine, or were a professional body such as that of the vignerons of France – had always been able to make a pronouncement regarding Pasteur's methods after just a few weeks of trials. As far as the impatient Pasteur was concerned, a year-long trial was unheard of and unnecessary.

Once Pasteur's letter had been read, Dr MacLaurin 'examined' Dr Frank Hinds, as the English-speaking member of the Pasteur delegation. Such an examination from the severe, unsmiling MacLaurin would have been more like a police interrogation. In his sing-song Scottish accent, the President began, 'I understand that you gentlemen come here as the representatives of Monsieur Pasteur, in regard to the claim which he has made for the reward of £25,000 offered by the government of New South Wales to the person who introduces a successful means of exterminating rabbits?'

'That is so,' said Hinds, who went on to say that the Pasteur Mission had not brought any written credentials other than the letter from Louis Pasteur which had just been read to the Commission, and letters of introduction to Premier Sir Henry Parkes and to the President of the Royal Society of New South Wales. He said that he and his colleagues were fully acquainted with the Pasteur scheme and were prepared to put it into operation in New South Wales.

'Have you been long associated with Monsieur Pasteur?' MacLaurin asked.

'I was only with Monsieur Pasteur for ten days before starting from Paris,' Hinds replied.

This answer would have raised eyebrows around the table, as would Hinds' next response to MacLaurin's question about his experience in micropathology laboratories. That experience, Hinds said, had not been extensive but rather, incidental. He hurried to add that he had worked with Professor Horsley at the Brown Institution and University College, London.[174] Professor Victor Horsley, later Sir Victor, was well known as a consulting surgeon with University College Hospital and as Superintendent of the Brown Animal Sanitory Institution in South West London, which specialised in research into diseases in domestic animals. In 1886, Horsley had headed a commission in England that had corroborated the effectiveness of Louis Pasteur's rabies vaccine. And just the previous year, Horsley had been the first surgeon in the world to successfully remove a spinal chord tumour from a human patient.

All this would have been mildly impressive to the members of the Rabbit Commission but they would have seen little relevance to the matter at hand – Horsley was not a microbiologist and he knew nothing about the application of chicken-cholera to rabbits. To them, Horsley's student Frank Hinds would have seemed little more than a messenger boy with a medical degree.

'Then, with regard to Monsieur Loir, has he been long with Monsieur Pasteur in matters of this nature?' MacLaurin continued.

'Yes,' Hinds replied. 'Some eight years.'

'In the study of this department of micropathology?'

'Yes.'

This was inaccurate – Loir had been working directly for Pasteur for the past six years, not eight. Six years or eight, this information should have sounded more reassuring, but the medical members of the Commission would remind their colleagues that despite his time with Pasteur, young Adrien Loir was still only a medical student.

'And Dr Germont?'

'He has been with Monsieur Pasteur for one year.'

MacLaurin seems to have been aware that Germont had actually been working with Pasteur's deputy Grancher in the rabies clinic for the past year, not directly for Pasteur, because he asked, 'With Monsieur Pasteur, in his laboratory?'

'Yes.' Hinds either lied or was ignorant of Germont's actual role in the Pasteur organisation.

MacLaurin moved on, asking if the team had brought a sufficient supply of chicken-cholera microbes with them. Hinds said they had, but until they conducted tests they would not know whether the microbes had survived the voyage. Dr Wilkinson now chimed in, suggesting that Pasteur's representatives be asked to prepare a comprehensive written statement on what Pasteur proposed for chicken-cholera experiments in New South Wales, for presentation to and discussion by the Commission the following day. Hinds agreed to prepare the statement and advised the Commission of the facilities the Pasteur Mission would require in order to carry out its work; he also asked for twenty rabbits for their initial experiments. MacLaurin then granted the Pasteur party permission to cultivate and test their chicken-cholera microbes in the laboratory.

Victorian commissioner Edward Lascelles was interested to know how the chicken-cholera would be communicated from one rabbit to another, asking Hinds whether he was aware that in the more barren districts of Australia, rabbits did not make burrows: in his written submissions, Pasteur had repeatedly said that infected rabbits would pass on the disease to other rabbits in their burrows. In response, Hinds told the Commission that he thought there would still be sufficient contact in open country, via the excreta of infected rabbits.

Dr Wilkinson then said, 'Some time ago we heard of there being some secret in the application of this method.' Pasteur had intimated as much in the letter just read. 'Is there any secret method of application which Monsieur Pasteur has not yet revealed?'

Hinds had no knowledge of any such a secret method but was wise enough not to contradict Pasteur. 'I will reply to that with the other questions.' He was referring to the requested written statement. In writing the statement he would seek Loir's guidance on this question of a secret method.

But New Zealander Fred Bell was not satisfied with this. 'But we all want to know whether such a method exists; we don't want to know the particulars.'

This reference to Pasteur's 'secret method' would emerge several times again in the future. In reality this 'secret method' was an invention of Pasteur's imagination. Pasteur was fearful that the Commission would simply acquire chicken-cholera microbes of their own, apply them to eradicating rabbits, and shut him out of the competition. The only secret was in Pasteur's method of attenuating the bacillus, involving the temperature used in the process. The actual application of the microbes to rabbits, both on their food and via inoculation, was quite straightforward. It appears that Pasteur was hopeful that reference to a secret method of application would discourage the Commission from attempting to acquire and use their own chicken-cholera microbes.

Frank Hinds held his line. 'I would rather wait and answer that question with the other questions.'[175]

Hinds, Loir and Germont then withdrew, hurrying to the Union Club to begin the struggle to commit to paper, in English, the statement required by the Commission. Despite the grilling they had just received, Adrien Loir went away enthused that the Commission may yet approve the Pasteur experiments. The fact that the Commission was asking for a written statement on Pasteur's chicken-cholera remedy belied the impression Loir had gained from his meeting with Frank Abigail. It seemed that the Pasteur contest entry was still in contention for the prize after all.

13

ENTER DOCTOR KATZ

AT 2.30 PM ON Monday 23 April, several hours after Loir and his companions had appeared before the Rabbit Commission, Dr Oscar Katz was called into the Executive Council Chamber to be interviewed by the commissioners. Pulled like a rabbit from a hat, Katz would prove to be a man who would, along with Dr Camac Wilkinson, be a key combatant in the battle that lay ahead between Pasteur's representatives and the Commission.

Oscar Katz was the commissioners' preferred candidate for the post of Chief Expert Officer. He was tall, with so little flesh on his bones that his clothes hung off him as though he were a clothes rack. His hair was dark and close-cropped, and he had a neat moustache and a goatee beard with a sharp, chiselled point. Spectacles perched on the end of his bony nose. Considerable mystery surrounds both the man and his Rabbit Commission appointment. Just as there is nothing on record to indicate that other candidates were considered for what would be an extremely well-paid post, not a great deal is known about Dr Katz. An Austrian, a native of Vienna, Katz appears to have been in his thirties when he arrived in Sydney in 1886. He spoke and wrote good English, if a little stilted, which suggests that he'd lived and perhaps studied for a time in Britain. In Sydney, it would be suggested that he had studied under Robert Koch in Germany. In Melbourne, where he would later briefly work, it came to be said that Katz had been a student of Karl Flugge, the German physician who wrote *Micro-Organisms with Special Reference to the Etiology of the Infective Diseases*.[176]

Soon after Katz arrived in Sydney in 1886 he was employed by the Health Board to conduct several bacteriological examinations of the water from the Sydney supply and, the following year, to make bacteriological observations at the Coast Hospital at Little Bay – the later Prince Henry Hospital. He was also granted membership of the Linnean Society of New South Wales, probably on the introduction of Health Board chairman Dr Normand MacLaurin who was a member of the society. The local branch of the Linnean Society, named after Swedish naturalist Carl Linnaeus, was founded in Sydney by wealthy philanthropist William Macleay to foster the study of natural history. Macleay, the society's first president, built Linnean House, a meeting hall and laboratory, on part of his own land in Ithaca Road, Elizabeth Bay.

When Oscar Katz appeared before the Rabbit Commission in April 1888, he gave his address as Linnean House, Elizabeth Bay. Linnean House did not have rooms like a gentleman's club. The only accommodation at the premises was a small, two-storey caretaker's wing. So Katz was either using the caretaker's accommodation or he was living elsewhere and gave Linnean House as his mailing address so that he did not reveal where he was living or with whom. If the latter, he was possibly living at the home of a Sydney member of the Rabbit Commission – either its president, Dr MacLaurin, or Dr Camac Wilkinson. This being the case Katz, a nervy, precise man, would have striven to avoid any accusation of nepotism in his Commission appointment.

There is a strong likelihood that Camac Wilkinson had prior involvement with the Austrian, having studied in both England and Vienna; perhaps the two had met in Europe, even studied together? Perhaps, after Wilkinson returned to Australia in 1885, he had written to his friend Katz to urge him to migrate to New South Wales, which Katz did the following year. It is hard to imagine Katz migrating to

Australia without encouragement from someone in the colony, for the number of German doctors then in New South Wales, or French doctors for that matter, could be counted on the fingers of one hand. Australia was not high on the list of attractive destinations for ambitious European medical men. It is not impossible then that Katz and Wilkinson were friends, and that Wilkinson had recently pushed for Katz's appointment to the post of the Commission's Chief Expert Officer. Then again, perhaps it was the Commission President, Dr MacLaurin, who was Katz's friend and promoter. Or maybe both doctors were on friendly terms with Katz.

Surprisingly, the usually comprehensive minutes of the Commission's proceedings merely say that, on the afternoon of 23 April, Katz 'was accorded an interview, and stated his qualifications for conducting such experiments as the Commission might require'.[177] Those qualifications were not recorded and, unusually, even though Katz met with the commissioners for close to two hours on this occasion, not a word uttered during this interview was entered in the minutes.

Two versions of this day's minutes appear in the official royal commission report. One, printed in the 'Minutes of Evidence' in the middle of the report, contains the verbatim account of the interview with Pasteur's representatives on the morning of 23 April but no minutes at all for the afternoon session with Katz. The other version, which appears earlier in the report in the 'Report of Proceedings' section, offers a summary of both the morning and afternoon sessions of 23 April, with just the brief mention of Katz's lengthy interview. The minutes of every other interview conducted by the Commission during the course of its sittings in three colonies were included in its final report. The omission of the Katz interview suggests that Secretary Mahon had been instructed by the Commission or its President to edit out this interview because details of Katz's qualifications may have proven embarrassingly inadequate to outside eyes.

Only a day after Katz's meeting with the Commission do we learn – from a motion in the Commission by Dr Stirling recommending Katz's appointment as Chief Expert Officer – that Katz was a PhD, a doctor of philosophy, not a doctor of medicine. That Katz was a PhD was later confirmed in the preface to the Commission's Progress Report. As a doctor of philosophy, Katz could have qualified in any area of the humanities or sciences. He may have trained in pharmacy, physics or chemistry, or even in metallurgy: his precise area of specialisation would never be revealed by the Rabbit Commission. Virtually every other researcher of note in the field of microbiology, including Robert Koch, was either a doctor of medicine or a veterinary scientist. All of Louis Pasteur's collaborators and assistants were doctors of medicine or like Adrien Loir, medical students. Louis Pasteur himself was not a doctor of medicine, but Pasteur was an exception in more ways than one. The revelation that Oscar Katz was a PhD does raise doubts about the extent of the Austrian's experience in the field of pure microbiological research.

On the morning of Tuesday 24 April, Oscar Katz was appointed the Commission's Chief Expert Officer at a salary of £60 a month – £720 over twelve months, an annual salary of $288,000 today. The Commission also made provision for the employment of an assistant for Katz. Based on a briefing from Commission Secretary Hugh Mahon late that same afternoon, the press would report: 'Dr Katz of the Linnean Society, an expert in bacteriology, was engaged by the Commission to superintend the experiments.'[178] Nothing that would appear in later Rabbit Commission reports would substantiate Mahon's claim to the press that Oscar Katz was 'an expert in bacteriology'.

On this Tuesday, Commission President MacLaurin again failed to attend the Commission sitting. Claiming pressure of other work, Dr MacLaurin would never attend another meeting of the full Commission. Months later, on 18 August, MacLaurin would officially tender his resignation from the Commission. He would never make another public utterance regarding the introduction of Pasteur's science into the colony, shrinking from the fight like a dying dog. Why would this man, who had been so anxious to keep Louis Pasteur's chicken-cholera bacillus out of Australia that he'd made two damning Health Board rulings against it, and who'd been so keen to sit on the Rabbit Commission, suddenly withdraw from his position of power? Even if he was overworked, it seems surprising that he would not have at least been able to attend some Commission sittings, particularly if his previously expressed fears over the introduction of chicken-cholera into Australia were genuine and heartfelt.

It is tempting to wonder whether Sir Henry Parkes did not discover a little dirt on Normand MacLaurin and discreetly threaten to disclose some sordid private secret unless MacLaurin ceased his opposition to Pasteur's science and resigned his government appointments. This suggestion may sound outrageous, yet within months MacLaurin would surrender all his government medical posts.[179] And, as you will see, cunning Sir Henry Parkes would soon spark an international incident over the Pasteur Mission with a behind-the-scenes manipulation of an equally sneaky kind, so such a tactic was not beyond him.

Parkes was to claim that throughout his career his only interest was the public interest. Parliamentary colleague William Walker would say that Parkes considered himself a tribune of the people[180] – in the Roman republic, the ten elected tribunes of the people could veto a vote of the Senate if they thought it was not in the public interest. Yet Parkes' personal motto was 'Caesar or nobody'. And Caesar was a dictator. As 1888 progressed, Parkes, feeling secure with his massive parliamentary majority, became increasingly more like a dictator, if a benign one, than a tribune.

Parkes did not nominate another New South Wales delegate to replace MacLaurin nor did he appoint a new Rabbit Commission president in his place. So, as a temporary replacement, the commissioners elected Professor Harry Allen to the post of Acting President. His presidency would become official once MacLaurin eventually resigned.

On 24 April, the new Acting President informed his colleagues that the Linnean Society had, snobbishly, advised that it could not permit persons who were not members of the Society to enter its premises – referring to both an assistant to Dr Katz and the members of the Pasteur Mission – ruling out the use of the Society's laboratory. But, said Allen, the government had now advised the Commission that Monsieur Pasteur's representatives were welcome to use the Health Board's Government Laboratory. This had eventuated as soon as Health Board President Dr MacLaurin had suddenly removed himself from the picture – even though MacLaurin would not officially resign for several months, he'd clearly ceased to play an active role in Health Board affairs from the moment he stepped down from the Rabbit Commission.

That afternoon Loir, Germont and Hinds again appeared before the commissioners and Hinds read aloud their prepared written statement. In it the team detailed Louis Pasteur's experience with the chicken-cholera bacillus, the experiments that Loir had carried out in France, and the experiments Pasteur had instructed Loir to conduct in Australia. The first two experiments Pasteur specified were small-scale, under controlled conditions. Pasteur's final chicken-cholera experiment was a large-scale inland trial on a sheep property currently infested with rabbits but involving a test area of no more than 500 acres.

After Hinds finished reading this document to the Commission, Professor Allen informed him that the Commission wished its new Chief Expert Officer, Dr Katz, to be involved in all the Pasteur team's experiments. The Commission had prepared a list of experiments that it wanted carried out, said Allen, and any experiments that Pasteur's representatives proposed could be performed 'but under the direct control of one of our officers'.

Hinds replied without hesitation. 'We will agree to that.'

'You then,' said Allen, 'associated with the other gentlemen who represent Monsieur Pasteur, will have the laboratory at your disposal, and Dr Katz will associate with you in carrying out these experiments.'

'Yes,' Hinds agreed.

Allen noted questioning looks on the faces of Loir and Germont, who had been trying to follow the conversation in English. So Allen said to Hinds, 'If you would like time to consult on this point, you can obtain it.'

Hinds did not as much as glance at Loir and Germont to receive their agreement. 'We accept your proposals at once,' he said, grateful that the Pasteur team was being given any chance to prove the effectiveness of Pasteur's submission to the Commission.[181] Hinds' rapid acceptance of the Commission's conditions would lead to major problems down the track.

Individual commissioners now began to throw questions at Hinds, questions that he answered relatively well. But when Professor Allen began to interrogate Hinds on Pasteur's lengthy 8 January letter, which Allen had clearly never read, Adrien Loir decided it was time that he spoke up. He politely corrected the professor and directed him to a section later in the letter where his question was answered by Pasteur. From Allen's comments it would seem that he was not even aware that this young Frenchman before him was the same Monsieur Loir who'd conducted the Paris and Pommery estate experiments referred to by Pasteur. As the questions continued, Germont joined Loir in providing answers. Between them, Loir and Germont parried numerous queries about chicken-cholera.

Eventually the questions turned to the effect of chicken-cholera on birds other than poultry, native birds in particular. Loir was aware that the competition's rules did not mention birds – an error on the part of Frank Abigail when drafting the judging criteria. As Professor Allen would point out to his fellow commissioners on 25 April, Abigail's proclamation had required that the winning method not be dangerous to horses, cattle, sheep, dogs and other useful animals, but birds were not mentioned. Legally, should Pasteur's method satisfy all the other competition conditions, yet killed every bird that flew, the prize should still be awarded to him. So Loir was perfectly candid with the Commission, commenting that he had witnessed two chicken-cholera outbreaks in France and had seen those outbreaks kill chickens, ducks, turkeys, pheasant and pigeons, but no geese or other form of wild bird.

South Australia's Dr Stirling then asked, 'Do you know the temperature here, how hot it sometimes is in Australia?'

'We do,' Loir replied. He had already pointed out, in the statement prepared for the Commission, that the chicken-cholera bacillus perished at 51 degrees centigrade (124 degrees Fahrenheit).

'And are you aware,' said Stirling, 'that it is often as hot as eighty-two centigrades in the sun?'[182] This was nonsense. The highest officially recorded temperature in Australia is 50.7 degrees centigrade (123.3 degrees F), recorded at Oodnadatta, South Australia, in 1960. In 1889, there would be an unofficial reading at Cloncurry, Queensland, of 53.1 degrees centigrade, (127.6 degrees F).[183] Perhaps a reading of eighty-two degrees came from a bushman's story, but it had no basis in fact. However, while they would have suspected that Stirling was exaggerating, neither Loir nor Germont argued with the South Australian commissioner.

'Yes,' said Germont, 'but it is not necessary to pick the hottest part of the day either to prepare or distribute the food [infected with chicken-cholera microbes].'[184]

A little later in the questioning, Camac Wilkinson asked, 'With regard to the septicaemia of the rabbit, referred to by Davaine and Koch – have you worked with it?'

'No,' said Germont.

Hinds remarked that Pasteur was still doubtful that chicken-cholera and rabbit septicaemia were alike.

'Koch has held that it is quite different,' said Wilkinson, before asking whether any member of the Pasteur team had worked with rabbit septicaemia.

Germont said that Loir had experience in this area, and Loir himself added that the ability of rabbit septicaemia to kill chickens was an unknown.[185] This diversion into rabbit septicaemia seemed pointless, and Commission President Allen changed the subject, pointing the interview toward a conclusion. But Camac Wilkinson's interest in both rabbit septicaemia and Robert Koch were to come into focus a little later when the matter of sabotage of the Pasteur contest entry raised its ugly head.

After several more questions directed at the Pasteur team, Professor Allen adjourned the sitting.

Early in the 24 April interview, Frank Hinds had informed the Commission that their microbes would not arrive in Sydney from Melbourne until 26 April. This was actually the date that the shipment from Noumea was due to arrive. Loir had telegraphed Melbourne with instructions for the chicken-cholera box being held there to be sent up to him by train, and it reached him on the 25th. While Loir and his companions waited for the 26th, when they would have both their Noumea and Melbourne microbes and be doubly armed for their experiments, the Rabbit Commission's newly appointed Experiment Committee went to work.

It was this committee's task to oversee all experiments carried out in relation to the rabbit competition. To begin with, all three New South Wales delegates proposed to sit on the committee because the experiments would be taking place in their colony. New Zealand's Fred Dillon Bell was also appointed to the committee, as he intended living in Sydney until the Commission's work was done. Victoria's Alfred Pearson then asked to join the committee. And Dr Joseph Bancroft of Queensland, realising that working on the Experiment Committee would put him in close contact with Pasteur's representatives and allow him to talk to them about helping the Queensland Government, also put up his hand.

In the end, the Experiment Committee consisted of Bancroft, Pearson, Wilkinson, Quin and Bell, and they elected Camac Wilkinson their chairman. Again, Premier Parkes would not have been pleased – as committee chairman, Wilkinson would write the research reports on which the Commission as a whole would base its determination regarding the rabbit prize. Wilkinson was now in a position to dictate how the competition panned out.

Wilkinson and his committee drew up specifications for a series of experiments for Dr Katz to conduct on their behalf. Some of those experiments would involve the so-called Tintinallogy Disease put forward by Doctors Ellis and Butcher, others were designed to test Professor Watson's rabbit scab. But the major focus would be on Pasteur's chicken-cholera microbes. Wilkinson told Oscar Katz that as soon as the Pasteur team had established the virulence of their microbes at the Government Laboratory, and as soon as a suitably secure test site was located, the Austrian was to commence the experiments that the committee had set down for chicken-cholera. For, via Frank Hinds' ready acceptance of Commission requirements, the Experiment Committee had gained the impression that the Pasteur team would meekly hand over its microbes and allow the Rabbit Commission's so-called expert to conduct all further tests to prove the effectiveness and safety of chicken-cholera as a rabbit extermination method.

Meanwhile, at the Union Club, Adrien Loir was waiting for the second microbe box to arrive on the 26th from Noumea. On the 25th, while he was filling in time in the Union Club lounge, practising his English, he was handed a short but intriguing letter from Gerrard Herring, Under Secretary for Stock with the Department of Mines. Herring wrote that his minister, Frank Abigail, who was also in charge of the Agriculture Department, wished to know if Loir 'would be prepared to teach someone in this Colony, to be named by the Government, the mode of cultivating and inoculating with the virus of anthrax as practised by M. Pasteur, and if so, upon what terms?'[186]

Loir was astounded. This request came from the same minister who only the week before had apparently condemned Louis Pasteur and his methods out of hand. How could someone so opposed to Pasteur have changed course 180 degrees, virtually overnight? Seeing a new Australian friend nearby, Loir called him over. Fifty-eight-year-old Walter Lamb, a former president of the Union Club, had befriended the young Frenchman the moment he arrived at the club, and in Loir's own words, 'took an interest in my progress'.[187] With a large property at Rooty Hill, Lamb was chairman of the Colonial Sugar Refining Company (CSR) and a director of the English Scottish and Australian (ESandA) Bank and of several insurance companies. Only months before, Lamb had opened a massive stone-fruit cannery in Sydney along lines he had seen in California, even bringing an American back with him to Australia to manage the plant.

Loir showed Lamb the letter from Under Secretary Herring, his disgust evident. Thinking that Abigail was playing with him, Loir told Lamb that he would not be going to see the minister. Lamb asked why not, and Loir told him about the previous week's meeting in which, he said, the minister had claimed Pasteur's method was worthless.

'That seems impossible,' said Lamb.[188] He knew Abigail, and was aware of his enthusiasm for his biological rabbit remedy competition. Lamb thought for a moment then said, 'Come with me.'

Walter Lamb took Loir several blocks from the Union Club to the Mines Department in Macquarie Street, then up the stairs to Secretary Abigail's office. Abigail welcomed them both, then threw back his head and roared with laughter when Lamb told him what Loir thought the minister had told him a week before. Now, between them, Abigail and Lamb explained what Abigail had actually said and Loir, to his acute embarrassment, realised how the Marquis De Rostain had misled him and caused him to walk out on the minister. Loir now told Abigail that one member of his party, Hinds, was talking about going back to England, and had almost convinced Germont and himself to do the same.

'Well then, I have to congratulate you for staying,' said Abigail. Then he got down to business. 'I cannot be of any help to your mission as far as rabbits are concerned because my policy has been defeated. But I would like to help you prove that my initiative has been of some value. There is a high death-rate amongst Australian livestock because of the Cumberland disease.' In fact, 300,000 sheep a year were dying from the disease in New South Wales, and a lesser number of cattle. On the worst-hit properties, the mortality rate was as high as forty percent of all stock. 'Anthrax is said to be the cause of that disease, but no proof exists,' said Abigail.[189]

This was the same issue that Sir Daniel Cooper's daughter had raised with Pasteur in Paris in February. But Pasteur's letter to Loir on this subject of Cumberland disease and anthrax had yet to reach him in Sydney. Loir told Abigail that it would not be as simple as just teaching someone in Australia to inoculate livestock with anthrax vaccine. It would first be necessary to conduct research to determine whether this Cumberland disease – named after the New South Wales county where it was first detected in 1847 – was indeed caused by the anthrax bacillus.

'Could you carry out some research on this?' Abigail asked hopefully. 'I will employ you at the government's expense.'[190]

Loir, sure that his uncle would not object to him testing this Cumberland disease to see whether it was in fact anthrax, offered to carry out initial experiments if Abigail could provide a sheep that a government veterinarian would certify had died from

Cumberland disease. Both Abigail and Loir were happy when Loir left the minister's office – Abigail because his political stocks would soar if he could cure the Cumberland disease problem, and Loir because he realised that by identifying the anthrax bacillus in Australian stock he would open up a huge new market for the Pasteur anthrax vaccine, and so make large sums for his uncle's Pasteur Institute fund.

Loir went immediately to the telegraph office. From all he had learned about Cumberland disease he felt sure that he would be able to prove that it was indeed anthrax. So, rather than waste time waiting until he had carried out tests on Cumberland disease specimens, he now sent a cable to Pasteur in Paris, asking him to send out a batch of anthrax vaccine so that he could trial it in New South Wales. Despite having to pay several pounds to send the telegram, Loir returned to the Union Club feeling pleased with himself, and grateful to his new friend Walter Lamb for taking the time and trouble to help him.

The initial meeting with Abigail, and the encounters with members of the Rabbit Commission, had soured Loir's opinion of Australians, but now the Frenchman was beginning to think that some of these colonials were damned fine fellows.

14

THE IMPASSE

ON FRIDAY 27 APRIL, the overnight Sydney Express from Melbourne drew into Granville Station and Premier Sir Henry Parkes nimbly boarded a First Class car. Inside, with the train rumbling into motion once again, Parkes joined his good friend Julian Salomons who was on the last leg of his return journey from England. As the train rattled in through the Sydney suburbs, Salomons filled the Premier in on political news from London, and told Parkes all about his February meeting with Louis Pasteur in Paris. Salomons assured the Premier that, from everything he had seen at the Pasteur laboratory, chicken-cholera truly was the answer to the colonies' rabbit plague, as Pasteur claimed.[191]

When the train pulled into Redfern Terminus shortly after, waiting to greet Salomons were a gaggle of journalists and more members of the Parkes cabinet – Mines Secretary Frank Abigail, Treasurer John Fitzgerald Burns, and Postmaster-General Charles Roberts. Telling the press only that his health had been improved by his European jaunt and that he would resume duties as government leader in the Upper House the following week, Salomons hurried away with the Premier and his cabinet colleagues.[192]

This same day, Adrien Loir made arrangements to begin his chicken-cholera experiments at the Government Laboratory in Sydney the following Monday, 30 April. One box of chicken-cholera microbes had arrived from Noumea and the other from Melbourne. It was now only a matter of testing the Noumea microbes to check if they were still active.

Since the meeting between Loir and Frank Abigail a few days earlier, the minister had wasted no time in taking up Loir's offer to carry out tests to see if Australia's Cumberland disease was indeed anthrax. Loir had subsequently received a message from Abigail: by next Monday or Tuesday a dead sheep would arrive in Sydney from Uarah Station, a property outside Matong – then called Devlin's Siding – in the Riverina district, sent by the station's owner, Arthur Devlin. The sheep carcass would come complete with written certification from both Devlin and Government Veterinarian Edward Stanley that it had died from Cumberland disease. Using this sheep, Loir would be able to carry out experiments to determine whether Cumberland disease was indeed anthrax. Loir would not tell the Rabbit Commission about any Cumberland disease experiments he might conduct at the Government Laboratory – it was none of their business.

On the Monday Loir, Germont and Hinds took their chicken-cholera microbes to the Government Laboratory in Albert Street, close to Bennelong Point – which today is the site of the Sydney Opera House but in Loir's day was home to a tram depot. Government Analyst William Hamlet had been in charge of the Government Laboratory for the past year. Thirty-eight-year-old Hamlet – short, with a closely trimmed beard and moustache, and speaking precise, cultivated English – had been inspired to study microbiology by the work of Pasteur and Koch. While research was not his chief interest – he would gain a glowing reputation for his forensic work in police cases – Hamlet eagerly made his facilities available to Pasteur's representatives.

With the newspapers of several colonies waiting with keen interest to report anything to do with Pasteur's mission, even though the journalists of the daily press were a little mystified about the actual scientific aspects of what was going on, at Hamlet's laboratory Loir fed three test rabbits food laced with the chicken-cholera broth from the Noumea sample. He already knew that the Melbourne microbes were deadly but he needed to verify the status of the Noumea batch. Besides, determined to avoid giving critics any ammunition they could use against the Pasteur Mission, Loir wanted to avoid admitting that he had illegally tested the contents of one microbe jar in Melbourne.

Days passed and the rabbits tested with the Noumea microbes failed to die. This was very quickly reported in the press as a setback for the Pasteur method. 'The reason assigned for the apparent failure,' said the *Sydney Morning Herald*, 'is attributed to the fact that the microbes had lost a certain amount of vitality during the long sea voyage and journey to Sydney.'[193]

Loir was not happy that this had been reported. He hadn't given the press or the Commission any information about his progress. Evidently, there was a 'leak' at the Government Laboratory, with someone on the staff passing information about the French experiments to the Commission, with Secretary Hugh Mahon subsequently briefing select friends in the press. It is likely that Mahon had cultivated a contact at the laboratory during his days at Sydney's *Daily Telegraph*. That contact is unlikely to have been Government Analyst William Hamlet, for he was a witness to the Cumberland disease experiments, and no news of those leaked out.

Loir, realising now that microbes from the Noumea box had either become weakened or had died – apparently after being stored in a hot place on the tropical island – turned to his Melbourne supply. By rights, his original microbes were still virulent, although he still did not reveal that he had already successfully tested them in the Victorian capital four weeks earlier.

Again an insider, although not a well-informed one, leaked details to the press. 'It is the intention of Monsieur Loir and his confrères to submit them to cultivation', the *Daily Telegraph* of 2 May reported of the second batch of French microbes, 'and to refresh or revivify those in which the virus is found to be deadened.' Next day, the *Telegraph* reported that three rabbits tested with the Melbourne batch of microbes had swiftly died. 'Monsieur Loir and his colleagues claim that this is a satisfactory proof of the efficacy of the remedy proposed by them,' said the *Telegraph*.

Meanwhile, by Tuesday 1 May, Loir had received his dead sheep from the Riverina. The carcass was delivered to the Government Laboratory where Government Analyst Hamlet, who was keen to see Loir carry out his Cumberland disease experiments, provided several healthy sheep and some mice in addition to the rabbits needed for the chicken-cholera experiments. Loir immediately inoculated two healthy young sheep with blood from the dead ewe. Within forty hours, one sheep was dead. At postmortem it displayed all the symptoms of anthrax. On 3 May, Loir inoculated four mice with blood from the dead sheep; three died within eighteen hours. Analysis under a microscope identified the anthrax bacillus in their blood. Loir, who was as familiar as any man on earth with the anthrax bacillus, had just proven that Cumberland disease was in reality anthrax.

This was the first significant microbiological discovery in Australia's history and Loir would have celebrated with his colleagues. The news was hurried to Frank Abigail. No doubt beaming from ear to ear, Abigail rushed to Sir Henry Parkes' office to tell him that Cumberland disease had been identified as anthrax – meaning that New South Wales could contemplate using Pasteur's anthrax vaccine to wipe out the disease in the colony, which would save many farmers a small fortune in stock losses.

Loir wanted to be sure he was right, so he used blood from the spleen of one of the dead mice to cultivate the guilty bacillus in gelatin and on potatoes. On the morning of 7 May, a rabbit was injected with this latest cultivation. By 3.00 pm the following day the rabbit was dead. In the presence of Government Analyst Hamlet and Government Veterinarian Stanley, Loir then conducted an autopsy and identified the anthrax bacillus in the rabbit's blood. That night, Loir wrote a report for Minister Abigail in his name and that of Germont and Hinds, at the end of which he formally concluded, 'Consequently, Cumberland disease and *charbon* (anthrax) are one and the same disease.'[194]

Abigail would quickly respond by suggesting that a trial should take place with the Pasteur anthrax vaccine on a typical sheep station – much like Pasteur's celebrated Pouilly-le-Fort trial of 1882 – to which all interested parties would be invited. If the trial proved a success, the minister would recommend government approval of the manufacture and use of the Pasteur vaccine in New South Wales. Loir immediately wrote to his uncle proposing that, if Pasteur approved, he and Germont conduct the anthrax trial requested by the New South Wales Government as soon as possible, and again asked for a supply of anthrax vaccine for such a purpose. Until Pasteur replied with his approval to proceed, and sent Loir the vaccine, nothing more could be done.

Although Loir had concluded his chicken-cholera experiments by 3 May, he would not submit a written report to the Experiment Committee on those experiments until 8 May, giving himself time to wrap up his Cumberland disease tests at the Government Laboratory. But on 3 May, the Experiment Committee learned via the informant at the Government Laboratory that the Pasteur chicken-cholera microbes were virulent and ready for full-scale experiments. Camac Wilkinson immediately pushed forward with plans for the next experimental stage while the microbes retained their virulence.

While waiting for Loir to conduct his laboratory experiments, the Experiment Committee had been searching for a suitable experiment site. The commissioners had agreed that this next stage must take place on an island, but after considering islands as far away as Bass Strait, between Victoria and Tasmania, it was realised that the expense and logistical difficulties involved in setting up and maintaining research facilities on a remote island would make it impractical. So they'd looked closer to home, considering locations in and around Sydney.

After it was pointed out to the Commission that Sydney's quarantine station was located on the mainland, the committee seriously considered several mainland sites including one near the government's railway workshops at Newtown which seemed ideal – until it was noticed that one of the city's largest poultry farms sat right next door; not an ideal place to test a disease fatal to chickens. The Quarantine Reserve at Randwick was thought to be perfect for the task, and Melbourne commissioner Albert Pearson also took a fancy to private property at Summer Hill. But as soon as the public learned that these sites were under consideration a flood of letters and a petition reached parliamentarians and the press from alarmed, misinformed members of the public who were terrified that Monsieur Pasteur's cholera germ would contaminate the surrounding air.

Under mounting pressure to keep the experiments away from the public, the committee had then examined islands in Sydney Harbour. 'We are hemmed in on all sides by those who fear danger,' Queensland commissioner Dr Joseph Bancroft told the Lands Department, according to the *Sydney Morning Herald* of 3 May. 'At the same time,' said Dr Bancroft, 'those who have made themselves acquainted with the subject think there is no danger.'

After inspection visits, Garden Island and Clark Island were both ruled out for various reasons, but Shark Island was felt to have potential. So, on the morning of 3 May, after learning that the Pasteur microbes were virulent, the committee adjourned and walked down the street to the Lands Department to obtain permission to use Shark Island for their experiments. They cornered Under Secretary Harrie Wood on the front steps of the Lands Department building but, as reporters gathered around with pencils and notebooks at the ready, Mr Wood proved evasive when committee members peppered him with questions about approval for an experimental site. Later that same day, the committee learnt from department insiders that Shark Island would not be made available because the government planned to use it as a quarantine station for imported cattle. A previously unnoticed island, Rodd Island, was, it was said, a more likely proposition, so that afternoon, Experiment Committee members Pearson and Quin took a government steam-launch out to Rodd Island to inspect it.

This little island of 0.6 of a hectare sits unobtrusively in Iron Cove between Rozelle and Drummoyne. Most Sydneysiders have never heard of Rodd Island but the influence of Louis Pasteur would make it temporarily famous in Australia during the last decades of the nineteenth century. Rodd Island had been set aside by the government in 1879 for public recreational use but no improvements had been made to it and it lay there neglected and deserted. The island that Pearson and Quin found was little more than a sandstone peak jutting from the water. The summit of that peak was flat and covered with scrub and a few trees, and was just large enough to accommodate a small research station. The only problem with the site was that quite a scramble was required to reach the peak.

Pearson and Quin reported back to Wilkinson, Bancroft and Bell that, with a little work, Rodd Island would be ideal for their needs. Pearson promptly took it on himself to approach a leading firm of Pitt Street architects, Sheerin and Hennessey, with a brief to design facilities for Rodd Island. The next day, 4 May, the Lands Department formally advised that Shark Island could not be provided, but offered Rodd Island in its place, and the Experiment Committee immediately wrote back accepting the offer. The government subsequently told the press that experiments on Rodd Island would not take many months, after which time the island would be returned to public use, while any facilities built there for the research could only be to the long-term benefit of the people of Sydney. This eased the transfer of the island to experimental use without public complaint.

The architects very quickly came up with a draft design and a cost estimate for the Rodd Island facility, but the Premier was not impressed by Pearson's initiative – the government had a perfectly capable Colonial Architect's Office for that sort of work. Pearson's draft design was passed onto Colonial Architect William Coles for plans to be drawn up and the cost of construction to be calculated. Premier Parkes would also require a change to Pearson's plan. The original plan provided for living quarters on the island for Dr Katz and an assistant. Parkes had a third bedroom added, allowing Loir, Germont and Hinds to live on the island rather than the Rabbit Commission's minions. The Premier was determined to do everything possible to encourage Pasteur's representatives to stay in the colony for an extended period and to put their skills to good use for New South Wales.

The design for the Rodd Island research station approved by Parkes provided a residential building, a laboratory building and a stables, all set around a quadrangle one hundred feet square that would be entirely covered with wire gauze. Adrien Loir was to joke that the Rabbit Commission designed the gauze enclosure to prevent chicken-cholera microbes from flying away. With so little known in Australia about the disease, the Experiment Committee's intent was to prevent birds and even flies from coming in contact with diseased rabbits in the enclosure, or their droppings, for fear of the disease being carried away and spread elsewhere. Alfred Pearson had remarked at a Commission sitting on 25 April, 'We should take into consideration the feelings of alarm expressed by the public, and we should take all possible precautions.'[195]

The Colonial Architect estimated that it would cost £2500 (a million dollars) to build the Rodd Island complex and its associated facilities, which would include a jetty, running water and a gas supply for the laboratory laid on via a pipeline from the mainland. Premier Parkes could see that once it had been built, the research station might provide long-lasting benefits to the colony, benefits that far outweighed the island's use as a public recreation reserve, and so he did not hesitate to authorise the expenditure. But just as construction work was about to begin on Rodd Island, a major obstacle to the testing of the Pasteur method emerged.

On 8 May, Frank Hinds wrote to the Experiment Committee, stating that preliminary experiments with the chicken-cholera microbes had been successfully completed and asking when the committee would be ready to examine Pasteur's scheme. The committee immediately wrote back, enclosing a list of experiments with chicken-cholera that it wished its Chief Expert Officer, Dr Katz, to carry out with the Pasteur microbes, and asking Hinds to suggest any other experiments that he and his associates might also want conducted.

When the Experiment Committee met again on the morning of Monday, 14 May, a letter from the Pasteur team dated 10 May awaited them. The letter was signed by Germont, Hinds and then Loir. There was a quite deliberate reason for this order of signing.

Experiment Committee chairman Camac Wilkinson had ignored Loir in meetings attended by the Pasteur representatives, directing his questions to Germont and Hinds – because they, like him, were qualified doctors and Loir was not. Loir, recognising this, from now on gave the impression that he occupied a back seat while Germont would seem to be taking the lead in discussions and negotiations. But it would always be Loir who, behind the scenes, was making the decisions on behalf of his uncle.

The French team's letter acknowledged receipt of the experiment schedule sent to them by the Experiment Committee. But then it said, 'We are unable to accept any participation in these investigations. M. Pasteur has instructed us to make certain experiments having for their object – strictly in accordance with the conditions for the reward published by the New South Wales Government – the demonstration that this method is at once efficacious, practical and without danger to the domestic animals mentioned in the note of the Government.' Pasteur's representatives said they were prepared to make those demonstrations that Pasteur had sent them to make, but could not participate in any other experiments that had no bearing on the competition.[196]

Loir was refusing to hand the Pasteur microbes over to the Commission and then stand back and let the Commission's so-called expert carry out experiments to prove or disprove Pasteur's claims for chicken-cholera. Camac Wilkinson and Alfred Pearson were incensed by what they saw as an unreasonable refusal to cooperate. They ignored the fact that Pasteur, or his representatives, always conducted their own experiments or trials when demonstrating Pasteur's methods. No doubt Pasteur supporters might wonder whether Wilkinson and Pearson would have expected Robert Koch to hand over his research to colonials to test if he, and not Louis Pasteur, had sent a team to Sydney. On the other hand, Rabbit Commission supporters would have argued that this was a formal competition, with a prize attached, and so the Commission was entitled to set the rules by which the winner was adjudged. Yet the scientific trials that Pasteur had submitted to in the past had also been competitions of a sort, each with a prize of glory and significant financial gain attached, and no one had attempted to take those trial experiments out of Pasteur's hands.

Both parties had legitimate arguments. Yet if the over-enthusiastic Mines Secretary Frank Abigail had spelled out in fine detail how the rabbit eradication competition was to be run and adjudged at the time of his 1887 proclamation in the international press, all parties would have known the rules that they were required to follow from the outset. This lack of detail had allowed both Pasteur and the Rabbit Commission to assume that it was with them that the burden lay for proving that the Pasteur method met the conditions for the awarding of the prize.

Wilkinson's fury when he read the letter of 10 May was generated by the apparent change of mind by the French. Three weeks earlier had not Frank Hinds, on behalf of the Pasteur Mission, agreed to cooperate with the Commission and allow it to conduct all experiments with the chicken-cholera microbes? The Commission President at the time had even asked if Hinds had wanted to confer with his colleagues before agreeing, but Hinds had consented without consultation and without hesitation.

The surprised and annoyed committee members were unanimous that they should require Pasteur's people to explain themselves in person. 'The attitude adopted by Monsieur Pasteur's representatives seemed to us so serious, and so difficult to understand after their statements to the Commission,' Wilkinson wrote in July, 'that we decided to call them before us without further delay.'[197] Secretary Mahon tracked the Pasteur representatives down, and at 2.30 that afternoon Germont, Loir and Hinds met with the Experiment Committee. When shown the minutes of the meeting where, according to the Experiment Committee, Frank Hinds had agreed to allow Oscar Katz to conduct all experiments with chicken-cholera, Germont and Loir shook their heads.

'We had no idea that such an arrangement had been accepted for us by Dr Hinds,' said Germont. Hinds must have been looking highly embarrassed at this point. Germont explained that the members of the Pasteur Mission had come to Sydney to perform a prescribed series of experiments set down for them by Louis

Pasteur, and they had no authority to deviate from Pasteur's plan. Nor were they permitted to hand their chicken-cholera microbes to the Commission until they had completed all of Pasteur's stipulated experiments.[198]

Seeing that the Frenchmen were adamant, the committee members reluctantly agreed to allow them to carry out the experiments prescribed by Pasteur. 'We wished, in fact, to allow them to put their own case in their own way,' Wilkinson wrote in July. But at the same time, the committee had no intention of altering its intention to also carry out its own experiments. In the tense atmosphere of confrontation, they asked the Pasteur team to provide them with a detailed schedule of the experiments Pasteur required. Loir and his companions agreed and departed.

Once Pasteur's representatives had left the room, the Experiment Committee concluded that Pasteur's experiments would, as far as they were concerned, be meaningless. They would humour the Frenchmen and allow them to carry out Pasteur's experiments merely to get their hands on the chicken-cholera microbes so they could conduct their own tests. It would only be based on the results of their own experiments that they would make recommendations to the Commission as a whole, recommendations to either accept that Pasteur's method of rabbit eradication was effective and safe, or to reject it.

There can be no doubt that, that night an unofficial meeting took place between the Commission's Chief Expert Officer Dr Oscar Katz and one or more members of the Experiment Committee. Alfred Pearson may have been present; he had become violently opposed to Pasteur's method. Dr Camac Wilkinson was certainly at the meeting. He would have been concerned that the Pasteur team might delay handing over their chicken-cholera microbes to the Commission, or not hand them over at all. At this meeting, Oscar Katz would have been asked where the Experiment Committee might get hold of chicken-cholera microbes of its own. There was anecdotal evidence, repeated by committee member Edward Quin at Commission sittings, of past chicken-cholera outbreaks in the colonies, but no one in Australia had identified or isolated the chicken-cholera bacillus. Such microbes would have to come from Europe, Katz would have replied.

Wilkinson now had an idea. He was familiar with a former member of Sydney's medical community and the local German community, Dr Carl Frank Fischer. Carl Fischer had taken his doctorate in medicine and surgery at Berlin University in 1857. In 1859 he had arrived in Auckland, New Zealand, where he operated a homeopathic clinic for several years before moving to Sydney in 1869. There, Dr Fischer had set up as a general practitioner and married a local woman some years his junior. Dr Fischer's home, Karlsruhe in Macleay Street, Potts Point, not far from Elizabeth Bay's Linnean House, had become a centre of the Sydney social scene. Since 1879, Fischer's wife, Sarah Jenny Fischer, had written the social news and women's pages for the *Sydney Morning Herald* and now for the *Sydney Mail.* For some time too, Dr Fischer had been estranged from his wife and their teenaged daughters, and had been living in Bavaria, Germany.

At Wilkinson's instigation, a telegram was sent to Dr Fischer in Germany, asking him to make contact with Louis Pasteur's bitter rival Robert Koch in Berlin, to request that Koch give him bacteriological samples to bring to Oscar Katz in Sydney. Fischer was to ask Koch for chicken-cholera microbes and rabbit septicaemia microbes – the subject of Camac Wilkinson's earlier dispute with Pasteur's representatives regarding their similarity or otherwise to chicken-cholera. The purpose of this request was to allow the Commission's man Oscar Katz to use these microbes in experiments designed to eliminate Pasteur's claim for the Australasian rabbit competition prize. It appears that Wilkinson offered to pay Dr Fischer's fare to Sydney from Berlin, from Commission funds, and perhaps also offered to pay him a fee for delivering the Koch microbes to the Rabbit Commission.

With chicken-cholera microbes supplied from Berlin by Robert Koch, Wilkinson and his Experiment Committee would not need Pasteur's microbes, and could conduct their own experiments no matter what the French did. It all now depended on whether Koch the famous bacteriologist, Wilkinson's idol, would help the Austrian

Oscar Katz in far-off Australia. Wilkinson would have been hoping that the carrot for Koch would be the knowledge that by helping Katz and Wilkinson he could sabotage Pasteur's attempt to win a major prize that would contribute to the creation of the Pasteur Institute in Paris.

The next day, 15 May, members of the Experiment Committee unexpectedly left Sydney, taking the train to Melbourne. Their excuse for suddenly leaving town was to attend a sitting of the full Rabbit Commission in the Victorian capital. In reality, that sitting was not scheduled until 21 May, six days away, prior to a sitting in Adelaide on the 23rd and further subsequent sittings around the colonies. The committee members' real reason for getting out of Sydney so quickly was to avoid receiving the requested list of Pasteur experiments that Adrien Loir and his two colleagues promptly drew up and which Frank Hinds would deliver to the Colonial Secretary's Department first thing on 15 May. This way, the Experiment Committee would not have to consider that list of Pasteur experiments until their return to Sydney in the middle of June.

The Experiment Committee was playing for time. And because all the committee members were participants in this stalling tactic, Wilkinson must have convinced every one of them to be a party to his plan, which was to delay the Pasteur team well into July, with Dr Carl Fischer hopefully arriving in Sydney by the first week of that month. If Robert Koch gave Fischer the requested microbes, Oscar Katz could be using them to conduct experiments in Sydney within seven weeks of Wilkinson's telegram being sent to Germany.

Once the Experiment Committee had their own chicken-cholera microbes, the Commission would be able to withdraw permission for Loir and his colleagues to experiment with their own microbes. As they saw it, this would allow the commissioners to regain the upper hand in their dealings with Pasteur. But ending the impasse with Pasteur's representatives all depended on whether Robert Koch was prepared to play dirty and sabotage Louis Pasteur by sending his own chicken-cholera microbes to Australia. Would Koch play along with the Rabbit Commission? Would he send his microbes to Sydney?

15

ROBERT KOCH'S SABOTAGE

AS MAY STRETCHED OUT and discussions with the Rabbit Commission dragged on, it became clear to Adrien Loir that Pasteur's hopes for a speedy decision on the contest prize were not going to be realised. With his temporary membership of Sydney's Union Club due to expire in the third week of May, and faced with the astronomical cost of an extended stay at the ritzy gentleman's club, Loir began looking around for cheaper accommodation for his colleagues and himself. The team's English 'interpreter' Frank Hinds finally proved of some practical use to Loir when he came up with a solution to the accommodation problem.

In making regular calls at the Colonial Secretary's Department over the past six weeks, Hinds had formed a good relationship with the Principal Under Secretary, Critchett Walker. The Principal Under Secretary was the head of the New South Wales civil service. In addition to being the Premier's right-hand man, Critchett Walker controlled eight different government departments, including the police and the colony's military. Walker was a career civil servant who was to serve a total of sixteen premiers of New South Wales during his twenty-four years in the top job. A sensitive man with a reserved yet courtly manner, and mechanically efficient in his work,

Critchett Walker was the glue that held the government of New South Wales together through its frequently changing administrations. Just short of his forty-seventh birthday at this time, Walker was a large man, and quite handsome, greying, with a moustache.

Walker was unmarried and never would wed – he was a confirmed bachelor, as it was termed in those days. This could mean that he simply never found the right woman to marry, but the usual implication was that he was homosexual. When Frank Hinds mentioned to Walker that he and his two French colleagues were looking for a flat to rent, Walker suggested that Hinds move in with him. This invitation encouraged Hinds to put on hold his plan to leave Sydney. His application for the job of hospital director had been unsuccessful – for the surprising reason that the hospital was looking for a man educated locally not in England, or so Hinds told Loir – yet Hinds did not go home as he had threatened to do.[199]

Critchett Walker was going to be away in Melbourne for six weeks, returning in early July. Victoria had cashed in on New South Wales' centenary celebrations by staging a Centennial Exhibition, a sort of trade fair, and New South Wales and the other Australasian colonies had set up stands at the exhibition. Under Secretary Walker was one of the New South Wales delegates appointed by Premier Parkes to attend the exhibition and officially represent the colony – the appointment amounted to a paid holiday, and was Parkes' way of rewarding Walker for his loyal service. Walker lived at Albany Chambers, an upmarket apartment building at 1 Bligh Street, just around the corner from the Union Club and a short walk from the Colonial Secretary's Department. He now offered Hinds the use of a bed in his flat while he was away, and longer if needs be. Walker also told Hinds that the two-bedroom flat right next door to his was vacant.

Loir acted quickly, and between 10 May and 14 May he and Germont moved out of the clubs where each of them was staying and into the vacant flat at Albany Chambers. At the same time, Frank Hinds moved from the Union Club and into Critchett Walker's apartment, occupying a bedroom that was linked to Walker's by a connecting door.[200] It was just after the trio had made this relocation that Loir learned that the Experiment Committee had left town, and that he could not expect another meeting with them to finalise the Pasteur experiments until the middle of June. At the same time, Loir was waiting for a letter from Pasteur authorising him to proceed with the anthrax field trial proposed by Frank Abigail.

With the Experiment Committee away for a month, and nothing to do but wait, Loir acquainted himself with Sydney. He bought a Sydney map that showed, drawn in perspective, every building in the downtown area and nearby eastern suburbs. On his map, Loir marked in French the principal places that now featured in his day-to-day life: the Colonial Secretary's Department where Sir Henry Parkes and Critchett Walker had their offices and, opposite it, the Lands Department where Loir received updates on the building work now being carried out on Rodd Island; the Botanical Garden where he would stroll; the lighthouse at South Head which he would have visited for its spectacular views of the ocean in one direction and back across the harbour to the city in the other. He also marked Kincoppal Sacred Heart School at Rose Bay. On the route to the South Head Lighthouse, the school opened its first permanent school building this year of 1888. On his map, Loir also indicated the location of the Union Club in Phillip Street, Albany Chambers in Bligh Street, the French Consulate then in Grosvenor Street,[201] a French café in George Street, and the French School in York Street.

There was a sizable French community in Sydney – upward of 2000 French citizens lived in the colony, many of them involved in trade between New South Wales and France. New South Wales had overtaken Victoria as the principal exporter of Australian wool, and France had become the second-largest buyer of that wool after Great Britain. In 1881, the French bank Comptoir National d'Escompte de Paris had opened a branch in Sydney to profit from that trade, and major French trading houses had wool-buying offices in Sydney – Leroux, Masurel, Caulliez and Mathon-Bertrand among them. French woolbuyers lived in some style in the city, and from 1889 they would organise well-attended Bastille Day picnics in Sydney every 14 July.

Yet Loir did his best to avoid the French community as he strove to speedily improve his English language skills and to make Australian contacts. The one private house he marked on his map was 'Maison Tooth', a waterfront mansion at exclusive Point Piper. The Tooth family were brewers and their Tooth KB Beer was among Sydney's most popular. Robert Tooth was a member and past committeeman of the Union Club, and Loir probably befriended him there – Tooth most likely sought him out after learning of Loir's experience with Carlsberg Beer and Victoria Bitter. Loir seems to have become a regular visitor at the Tooth house.

Despite befriending the likes of Walter Lamb and Robert Tooth and being welcomed into Sydney homes, this was a frustrating time for Loir, not only waiting for the Rabbit Commission to reconvene in Sydney but knowing that his letter to Pasteur seeking approval to proceed with anthrax trials would take thirty-seven days to reach Paris, with replies from his uncle taking another thirty-seven days to come back from the other side of the globe.

Sir Henry Parkes, meanwhile, was keeping a close watch on the Rabbit Commission, and shaking his head at the way it appeared to be deliberately stalling Louis Pasteur's representatives. Parkes pressed for construction of the research station on Rodd Island to be sped up – a team of forty workmen on the island was encountering problems, and the completion time had blown out from two weeks to some time in June. First, the solid sandstone on the island had proven an obstacle, so building supervisor Purkis had sought and received permission to blast the rock to make a road up the little peak from the shore to the summit. Later it was found that the gas pipe laid from the mainland didn't work, so a gas supply had to be created on the island using a petrol generator. But at least Sir Henry found Frank Abigail's anthrax developments encouragingly positive.

Those developments had taken another forward step. The one helpful thing that the Rabbit Commission had done for Pasteur's representatives was arrange for twelve-month First Class rail passes for all three of them. This meant that Loir, Germont and Hinds had only to show the small red leather rail pass wallets to travel on any train anywhere in New South Wales free of charge. Around the beginning of the last week of May, with the Experiment Committee out of town, Loir, Germont and Hinds joined Government Veterinarian Edward Stanley to ride a train into the Riverina district to pay a visit to Arthur Devlin's Uarah Station. Whether this trek southwest to the sweeping alluvial plains beside the Murrumbidgee River had been conceived by one of the bored members of the Pasteur team or whether it was Edward Stanley who initiated it, we cannot know, but it was potentially to the advantage of both the Pasteur Mission and the Parkes Government for them to make the journey.

At Devlin's Siding the party was met by Arthur Devlin, who drove them in his buggy out onto sprawling, flat but relatively green Uarah Station. Devlin was delighted to see the scientists; Cumberland disease had him at his wits' end, for the disease had been ravaging the Riverina for eleven years and, according to Arthur Devlin's descendants, he lost 80,000 of his 120,000 sheep to it in just four years.[202] When Devlin showed his visitors several sheep that had recently died, Loir and Germont took blood samples. Producing his microscope, Loir once again confirmed on the spot that Cumberland disease was indeed anthrax. Leaving Devlin hopeful that the French microbiologists might solve his disease problems and in the process help him make his property a paying proposition once more, the party returned to Sydney.

At William Hamlet's Government Laboratory, Loir produced a culture of the anthrax bacillus using the blood from the dead Uarah Station sheep. He then gave the culture to Government Veterinarian Stanley, telling him that once they had approval to proceed from both the New South Wales Government and Pasteur, and once the vaccine arrived from France, they would use that culture to infect test animals in the proposed anthrax trial.[203] By early June Loir, back in Sydney, was cabling his uncle to tell him about the productive trip to Uarah Station, and to urge him to send anthrax vaccine as soon as possible.

On 25 May, around the time that Adrien Loir and his companions were heading for Uarah Station, New South Wales Agent-General in London Sir Daniel Cooper was approaching the British Government to try to ascertain what arrangements Louis Pasteur had with foreign governments for the sale of his anthrax vaccine. This information was being sought by Sir Henry Parkes to give his government a basis for negotiating an anthrax vaccine contract with Pasteur should the proposed Uarah trials prove successful. It was Cooper's understanding that Pasteur had an arrangement of this kind with the Spanish Government but he asked Downing Street to establish the facts for him.

The British Foreign Office passed Cooper's inquiry onto the British ambassador in Paris, Lord Lytton – Edward Bulwer-Lytton, son of the famous novelist, the elder Lord Lytton. Himself a well-regarded poet, the younger Lord Lytton, previously Viceroy of India, was not only fluent in French but was on speaking terms with Louis Pasteur; he went directly to Pasteur for the information Cooper was seeking on Parkes' behalf. Pasteur's response was to cause surprise and consternation in both London and Sydney.

In answer to the question about the sale of his anthrax vaccine, Pasteur told Lord Lytton in a 11 June letter that he had a commercial agreement with the government of India, but in other countries including Spain he dealt through private agents. Pasteur also took the opportunity to promote his proposal for the eradication of rabbits in Australia using chicken-cholera. 'I repeat here that my procedure is absolutely harmless for domestic animals, birds and man. I don't speak lightly. This is a result of my personal observations and experiences.'[204] Pasteur then went on to complain about the way his delegates were being treated in New South Wales, expressing 'my profound regret that my delegates have found since their arrival in Sydney, contrary to all the provisions [of the advertised contest], hostility which makes it in return very difficult to accomplish their mission if the Sydney government doesn't make them welcome'.[205]

Lord Lytton immediately sent Pasteur's letter to Robert Cecil, Marquis of Salisbury, who was both Prime Minister and Foreign Secretary of Great Britain at that time. Lytton noted, of Pasteur and his delegates in New South Wales, 'He states that unless the government affords them all its support it will be difficult for them to accomplish their mission.'[206]

Lord Salisbury was a Tory who, unlike his Liberal predecessor Gladstone, was not on the same side of the political fence as Sir Henry Parkes. And unlike Gladstone, he did not particularly like Parkes. Pasteur's unfavourable reflection on the Parkes Government did not well please the British Prime Minister, or endear him to the New South Wales Government, for whose foreign policy Salisbury was responsible. On 19 June, Salisbury's deputy at the Foreign Office, Sir James Fergusson, a former governor of both South Australia and New Zealand, wrote frostily to the Colonial Office on Salisbury's instructions, to draw the Colonial Secretary's attention to Pasteur's complaint. After all, Fergusson wrote indignantly, Pasteur's delegates 'are there for the purpose of instructing the government in his method for the destruction of rabbits'.[207]

A week later Lord Knutsford, the Principal Under Secretary of State for the Colonies, passed on the complaint to the Governor of New South Wales, Lord Carrington, along with the British Government's huffy disdain for the way that Pasteur's people had been treated in the colony.[208] No doubt in the corridors of Whitehall, such treatment would seem typical of unpolished colonials. No one had told the Prime Minister or Fergusson or Knutsford that Pasteur was engaged in a competition for a £25,000 prize in New South Wales, and was not merely 'instructing' the locals as Fergusson had written.

Once Lord Carrington received this letter from London in August he would raise it with Sir Henry Parkes when he next came to dinner at Government House. In response, Carrington, who was extremely fond of Sir Henry and treated him with filial respect, received the Premier's assurance that this affair was nothing but a storm in a teacup. The Governor said no more about the matter. But this was not to be the last time that Pasteur's bid for the rabbit prize would spark an international incident. The next time it would even embroil the Prince of Wales.

On Monday 11 June, the Rabbit Commission's Experiment Committee met again in Sydney after a four-week hiatus and Loir, Germont and Hinds immediately asked for an interview, which was granted the next day. Because Camac Wilkinson was late for this 12 June meeting, Victoria's Alfred Pearson took the chair.

When the members of the Pasteur Mission sat down in front of the committee, Pearson, instead of referring to the team's letter of 14 May in which the French team had set out the planned Pasteur experiments as the committee had requested, began by asking Germont if he had any objections to the committee undertaking an experiment similar to one carried out by Loir in Paris, mingling uninfected rabbits with those infected with chicken-cholera.

Germont must have sighed before replying. The Pasteur team, he said, had been sent to conduct experiments set down for them by Pasteur, and they had no authorisation to change that. He reminded the committee that the Commission could conduct whatever experiments it wanted once the Pasteur experiments had been completed. 'When Mr Abigail, with the approbation of the government, proposed a prize for the best method of destroying rabbits,' Germont went on in his laboured, heavily accented French, 'he fixed the conditions and programme for the prize'.[209] That was that, as far as the French were concerned. Even if the Commission wanted to change its instructions, Germont added, instructions that took the form of the conditions set down by Minister Abigail, the Pasteur team had no authority to change the instructions it had received from Louis Pasteur.

Pearson then asked if the French would hand over samples of their microbes after they had completed the first of the three experiments that Pasteur required them to carry out. Naturally, Germont's reply was in the negative. Germont then said that all the Mission could do was write to Paris for further instructions. This caused the committee members to go into a huddle for a private discussion. After this, Pearson proposed to the French that they send a telegram to Pasteur, at the Commission's expense, seeking approval for a change in experiment conditions. Germont glanced at Loir, who nodded. Germont agreed to the proposal. Both sides then adjourned to work separately on the wording of the telegram.

Next morning, the parties met again, and the wording for a telegram to Pasteur was agreed: 'The Commission cannot allow your experiment to be made on a large scale before contagion from rabbit to rabbit has been proved. Even after that it cannot promise beforehand to make experiments on a large scale. Do you authorise the following experiment in contagion?' Details of the proposed experiment followed, after which the telegram concluded, 'The Commission also asks for infected rabbits for special experiment.' With the wording agreed, Germont translated it into French.[210] The draft telegram, which would go in Germont's name, was taken by Secretary Mahon to be sent by the Colonial Secretary's Department, and the meeting adjourned. The telegram was promptly approved by Premier Parkes and transmitted to Paris via Agent-General Sir Daniel Cooper in London.

Two and a half months had now passed since the Pasteur team had landed in Australia, and not a single experiment had been conducted with chicken-cholera microbes other than those designed to prove their virulence. So much for Pasteur's desire for all the experiments to be over and done and the prize handed over within four to six weeks. According to Pasteur's original schedule, Adrien Loir should have been walking back in the door at the Rue d'Ulm about now bearing the 625,000 franc prize.

On 16 June, Pasteur cabled his reply to the Commission's telegram: 'Contagion great in the burrows. Make experiments demanded on solid ground of a square metre. Wait before giving microbes of dead rabbits. But give drawings of the organisms of the cultivations, to be signed by the members.'[211]

At this point, Pasteur was not happy about making concessions to the Rabbit Commission, and annoyed that they persisted in their demand for proof that chicken-cholera was contagious among rabbits. Within a day of sending this latest telegram to Loir, Pasteur was complaining about the Rabbit Commission's demands in a letter to a friend in England, Mrs Eliza Priestley. She was an avid vaccinist whose husband, William Overend Priestley, was a leading London surgeon, and the Priestleys shared several friends and acquaintances with Pasteur including American physician and author Oliver Wendell Holmes.

'Why put so much emphasis on contagion?' Pasteur wrote to

Mrs Priestley on 17 June. 'All the Australian burrows are not linked and it will be necessary to proceed burrow after burrow, field after field.' [212] Pasteur, obsessed with winning the rabbit prize, was now haunted by this Antipodean doubt about the contagiousness of his rabbit remedy. The day after he wrote to Mrs Priestley, he was writing to Julian Salomons, the Parkes Government minister who had visited him in February. Salomons had impressed Pasteur, who saw the Australian lawyer and parliamentarian as a potential ally in a sea of Antipodean thieves, and now he confided in him: 'I know, now, how the contagion occurs, but the contagion question is of minor importance.'[213]

In Sydney, at the rabbit competition battlefront, as soon as Loir and his colleagues received Pasteur's telegram of the 16th they took it to a meeting of the Experiment Committee. There, Germont translated it for the locals. When Wilkinson and his fellow committee members said they were unclear about what Pasteur had actually said, Germont explained what he and Loir believed Pasteur's instructions to mean. 'We will give proof to the Commission that for each of our experiments we use the same microbe and same cultivations. And for this purpose the expert named by the Commission shall examine, with microscopic preparations of our cultures, and make drawings of the microbe which he shall sign as a guarantee of their correctness.'[214]

Wilkinson thought that this experiment would only be a waste of time. He demanded to know if the Pasteur team would hand over their microbes once they had successfully conducted it.

With Wilkinson again seeming to move the figurative goalposts of the competition, Germont's patience was running out. 'You have sent a special dispatch to Monsieur Pasteur,' the Frenchman responded, 'who has made a direct reply. We cannot modify the instructions which he has given us.' It was useless to ask Pasteur for further instructions on this new question, Germont said, adding: 'We can only make the experiments which we are authorised to make.'[215]

Now Fred Bell from New Zealand asked if the Frenchmen would conduct several variants on the experiment that Pasteur had just authorised – using six two-metre enclosures.

Germont and Loir had had enough: 'No, we refuse to do that,' declared Germont before he, Loir and Hinds came to their feet and unceremoniously trooped from the room. [216]

Once the members of the Pasteur Mission had left, the committee members looked at each other, with several of them uncertain what to do next. Then Fred Bell proposed that they compromise and authorise the French to conduct the experiment that Pasteur had agreed to in his telegram on Rodd Island, commencing on the following Monday. But Alfred Pearson was against giving ground. As Bell and Dr Bancroft began to argue with him, Pearson threw his hands in the air, jumped up, and angrily stormed from the room. In his absence, his colleagues agreed to Bell's proposal. But that proposal was never communicated to the Pasteur team. As the events of the next few days were to show, Pearson was later able to convince the other members of the committee that, rather than give in to the French, they should do everything in their power to retain the upper hand.

As soon as Loir left this unsatisfactory meeting with the Experiment Committee he sent a telegram to Pasteur, again at his own expense, to tell him that negotiations had broken down. At the same time, he warned, 'Systematic and combined attempts have been made to prevent Pasteur delegates receive fair play. Commission will be influenced to demand absurd demonstrations.' [217]

On 19 June, Premier Parkes received a telegram from his London Agent-General, Sir Daniel Cooper: 'Confidential. Pasteur urgently requests you will allow him telegraph free to Sydney and receive replies free. Necessary for him communicate frequently about Rabbit Commission matters.' Pasteur had apparently never overcome the shock of the cost of the telegram he had sent to Frank Abigail in January. In this latest telegram, Pasteur also went on to complain of unfair treatment by the Rabbit Commission before saying: 'Salomons understands subject. Appoint him *amiens curial* [a friend in court] to protect delegates.' Julian Salomons had certainly left an impression on Pasteur as an honest and upright man. Pasteur also asked Cooper to tell the Commission, 'No danger in trial on rabbits in the open. It is done in France without restriction.' [218]

Premier Parkes read this telegram and then filed it. He would have been totally unimpressed by the conflict between the Rabbit Commission and the Pasteur Mission, but for the moment he refrained from becoming personally involved. He didn't send Julian Salomons to the Commission to act as the legal representative of Pasteur's delegates as Pasteur had requested, nor did he intercede in the Commission's proceedings. Instead, the Premier not only kept this telegram to himself, he refrained from informing either the Commission or Pasteur's delegates of its contents. But he did give serious consideration to Pasteur's request for the cost of telegrams to be covered by the New South Wales Government.

The Experiment Committee reconvened on Monday 18 June, with the obstreperous Alfred Pearson once again taking his seat with the other committee members. Over the next three days the committee met daily for sometimes heated discussions. It seems that Wilkinson and Pearson were all for taking a hard line with the French while Quin, Bell and Bancroft had a more conciliatory attitude. Queenslander Bancroft, in particular, would not have wanted to upset Pasteur's representatives, for these were men he had been sent south to coax to Queensland. Yet none of what was actually said during the Experiment Committee meetings over these three days was recorded in the minutes of the later Commission report, unlike every other meeting of the committee. All the minutes for the 18–20 June meetings were to say was, 'The committee deliberated.'[219] Clearly, their deliberations were heated.

Experiment Committee chairman Camac Wilkinson, meanwhile, was acting as though he were the Rabbit Commission personified. On his own authority he had employed a cook and a handyman for Rodd Island at Commission expense, without any authority from the Commission as a whole to do so, and had written to Sir Henry Parkes seeking expense money for the members of the Experiment Committee – an act which the full Commission would later have to condone. Now, in the face of the intransigence of Loir and Germont, Wilkinson was determined to show the French who was running this rabbit competition. In the end, he and Pearson won the argument and convinced their colleagues to back a united and unwavering stance.

By the afternoon of the 20th, the Experiment Committee had hammered out what Wilkinson was to accurately describe as an 'ultimatum' to the French.[220] On 21 June, the Experiment Committee presented this ultimatum to a full sitting of the Rabbit Commission, seeking its approval.

'We felt that our only duty was to leave all further action in the hands of the full Commission,' Wilkinson told the other members of the Commission, at the same time characterising the experiment proposed by Pasteur in response to the Experiment Committee's telegram as 'of little value'.[221]

The ultimatum that Wilkinson and the Experiment Committee put before the full Commission was this: Unless Pasteur's representatives agreed without delay to provide rabbits infected with chicken-cholera for experiments conducted under the direction of the Commission, then correspondence with Pasteur and his representatives would be suspended, previous permission given to Pasteur's representatives to possess and experiment with chicken-cholera microbes would be withdrawn, and the Commission would authorise that 'steps be taken to obtain the microbes of chicken-cholera from other sources'.[222]

It appears that most other Rabbit Commission members were at this stage unaware that five weeks earlier Wilkinson had, via Carl Fischer, approached Robert Koch in Berlin, for an alternative supply of microbes. As proof of this, the following day, 22 June, South Australian commissioner Edward Stirling moved a motion in the Commission proposing that, if Pasteur's representatives did not conform to the Commission's ultimatum, then the Experiment Committee was to 'be instructed to make arrangements for getting a supply of the microbes from Europe'. Stirling's motion was endorsed by the Commission without a dissenting vote.[223] This motion flew in the face of the fact that, as Wilkinson and possibly other commissioners including Pearson knew – providing Koch had agreed to Wilkinson's May request – those microbes would already be on the high seas and heading for Sydney in the care of Dr Carl Fischer.

On 21 June, all of Wilkinson's fellow commissioners voted in favour of sending the Experiment Committee's ultimatum to the Pasteur Mission at once. With one voice they declared, 'The Commission expresses its surprise that as Monsieur Pasteur is seeking to obtain a reward of £25,000 for his scheme, his representatives should have absolutely declined to permit the fullest testing of its merits.'[224]

A written copy of the Rabbit Commission ultimatum was hand-delivered to Adrien Loir the same day that it was approved by the full Commission. As soon as he received this, Loir wrote to his uncle, enclosing his translation of the ultimatum. He also sent Pasteur a telegram, briefly telling him of the ultimatum's contents.[225] There was no immediate reply. Pasteur, it seemed, was both taken aback by the Commission's stance and reluctant to incur the expense of sending a replying telegram.

On 22 June, planning to conduct more hearings in rural areas, the Rabbit Commission adjourned. Its next Sydney sitting was not scheduled until 18 August. This same day, 22 June, Premier Parkes sent a telegram to Agent-General Cooper in London: 'I have no objection to defray telegrams Pasteur referred to in your telegram of 19th June.'[226] This agreement by Parkes to cover the cost of telegrams from Pasteur and from Loir and Germont would have been passed on to Pasteur in Paris the next day, a Saturday.

On the following Monday, 25 June, Pasteur responded to Loir's last cable, knowing that his telegrams and replies from Loir and Germont would not from then on cost him or Loir a single franc. 'Give cultures for experiments according Programme Prize 31 August 87,' his telegram said, 'stipulating for your presence at all experiments with the cultures but refuse absolutely all techniques for preparation of cultures.' Pasteur had also by this time received Loir's first letters from Australia, in which Loir had mentioned the Commission's concern about chicken-cholera spreading to wild birds. So Pasteur added, 'I affirm there is no danger to the birds at liberty.'[227]

In this same telegram Pasteur addressed the Commission's doubts about contagion among rabbits infected with chicken-cholera: 'I have made experiments with five infected and twenty healthy rabbits in three days,' he said. This was the main reason for his delay in replying to Loir's telegram: as soon as he received it, Pasteur had hurried to the Rue d'Ulm laboratory to carry out three days of additional chicken-cholera experiments. 'The five infected and eleven by contagion died,' he advised.[228] This, to Pasteur, was positive proof that chicken-cholera was contagious and self-propagating among rabbits.

After Loir read this telegram, he deliberately went against his uncle's wishes. From his point of view, here was Pasteur, still in poor health after his stroke the previous October and desperate to win the prize money, allowing himself to be pressured by the Rabbit Commission into giving up control of his own experiments – for the first time in his career. Loir dug in his heels and decided not to pass on his uncle's telegram to the Rabbit Commission, or to even tell them about it. He did this not for any reason of ego or personal ambition, but because of his loyalty to Pasteur and his desire to do what was best for him. In his heart, Loir knew that Pasteur would forgive him, and even thank him in the end. Because above all he knew that Pasteur trusted him, implicitly.

It had not always been the case. Four years before, Pasteur had accused Loir of making a mistake in laboratory work with swine erysipelas. 'Oh, Lord! Oh, Lord!' Pasteur had cried, pacing up and down. 'Adrien, what have you done? Why did I trust you?'[229] Loir had been heartbroken that his uncle had doubted him, but he was certain that he'd done the right thing, and swore to Pasteur that he hadn't made a mistake. Pasteur did not believe him until after Loir conducted an autopsy on a dead pig at the Côte du Nord at 4.00 in the morning. From the Côte du Nord, he had sent slides of blood taken from the pig back to Paris. The next day, Emile Roux had cabled Loir to say that the young man had been right all along. When Loir returned to the Rue d'Ulm, an emotional Louis Pasteur embraced him. 'He rarely embraced anyone,' Loir wrote later, 'so this gesture showed that I had regained his confidence.'[230]

Pasteur had never doubted Loir since then. And young Loir, so deeply loyal to Pasteur, was certain that once again his uncle was wrong and he was right. Determined not to reveal that Pasteur had backed down in the face of the Rabbit Commission's ultimatum, Loir determined to take decisive action of his own. He waited until the following week when Principal Under Secretary Critchett Walker returned to Sydney from his six-week stint at the Centennial Exhibition in Melbourne. Frank Hinds was still living in Walker's flat and would cohabit with Walker until the end of August.[231] Either Loir personally briefed Walker on the problems he was having with the Rabbit Commission, and of the action he proposed to take in response to those problems, or Hinds informed Walker on Loir's instructions. Either way, by involving the Principal Under Secretary, Loir achieved his desired reaction.

The next day, Critchett Walker reported to Premier Parkes that the Pasteur delegates had taken all they were going to take from the Rabbit Commission, and that Loir was refusing to have anything further to do with the Commission. Now Parkes acted. He sent for Camac Wilkinson MP. When Wilkinson appeared in the Premier's office, Parkes informed him that, 'Monsieur Pasteur's representatives had expressed a wish to sever their connection with the Commission', and then proceeded to dress Wilkinson down as though he were an errant schoolboy. Trying to defend himself, Wilkinson blustered that the Commission was in the right and Pasteur's people in the wrong, and he offered to write a memorandum for Parkes setting out all that had transpired between the two parties to justify the Commission's 21 June ultimatum.[232]

As far as Parkes was concerned Wilkinson was free to submit a memo to him on any subject that took his fancy, but the Premier was not going to tolerate any more nonsense from the Rabbit Commission. He informed Wilkinson that Monsieur Pasteur's representatives would be permitted by the Lands Department to undertake the experiments that Pasteur had sent them to New South Wales to conduct, on Rodd Island, beginning the following Monday. He also reminded Wilkinson that Pasteur had agreed that once they'd completed his experiments, his delegates would provide chicken-cholera microbes to allow the Commission to undertake whatever further experiments it wished; Sir Henry had no reason to believe that Pasteur would not adhere to this undertaking. And as for the ultimatum from the Rabbit Commission to Pasteur, it was null and void as far as the Premier was concerned. Parkes then curtly dismissed the arrogant young member for Glebe from his presence.

Wilkinson was humiliated. Even though the Rabbit Commission had been acting as though it controlled Rodd Island, the island came under the jurisdiction of the Lands Department, and the Premier had every right to rule who could or could not use it, and when. Realising that he had absolutely no power to stop Pasteur's people conducting their experiments on the island, Wilkinson didn't bother to convene the Experiment Committee or to contact other Commission members. He informed Dr Oscar Katz that the Pasteur representatives would be commencing their experiments on Rodd Island on Monday next, 7 July, and instructed him to be there to watch their every move.

On the morning of 7 July, Loir, Germont and Hinds boarded a Government steam-launch at Circular Quay. This transferred them to Iron Cove and landed them at the brand new jetty at Rodd Island. Loir and Germont had their personal belongings with them along with their chicken-cholera containers, for they would be living on the island while they conducted the Pasteur experiments. Dr Katz, the Rabbit Commission's Chief Expert Officer, would be occupying the third bedroom in the island's accommodation wing so that he could observe and record all the experiments undertaken by the French team. Frank Hinds would continue to live with Critchett Walker at Albany Chambers. In practice, Loir and Germont would be conducting the experiments. Hinds was to have little input, spending much of his time during the coming weeks at Albany Chambers, only occasionally bothering to go out to the island.

On that same morning of 7 July, Dr Oscar Katz also prepared to catch a boat for the half-hour trip to Rodd Island from downtown Sydney. But he was delayed by a visit from Dr Carl Fischer, who had just arrived in Sydney from Europe and had something for him. Fischer gave Katz several containers from Robert Koch in Berlin – one contained chicken-cholera microbes; the other, rabbit septicaemia microbes. Louis Pasteur's great German rival Koch had indeed agreed to the May request from Camac Wilkinson and Oscar Katz, and was actively participating in an attempt to sabotage Pasteur's rabbit competition entry.[233] The Experiment Committee could now conduct their own experiments, and sideline the Pasteur team.

16

SIR HENRY'S COVERT ACTION

EVERY PIECE OF MAIL addressed to Premier Parkes was delivered to the Clerk of Records in the Colonial Secretary's Department. Around the middle of the day on Tuesday 10 July, a small wooden box and an envelope that had been landed from the French steamer *Oceanien* that morning were delivered to the desk of the Clerk of Records. It was the clerk's job to open all correspondence, record it, determine its contents, then distribute it to the appropriate officer who would be in a position to deal with it – very little of the thousands of mail items addressed to the Colonial Secretary every year actually reached the Premier's desk.

The clerk looked at this box and envelope uncertainly. Both were addressed to 'Sir Henri Parkes, Premier du Government, Sydney'. The clerk opened the envelope to find that it contained a note, dated 26 May, from Louis Pasteur to Premier Parkes, as well as a second, smaller sealed envelope addressed to Messieurs Loir and Germont and marked 'Confidential'. The note to the Premier said, in poor English: 'I have the honour to remit to you under this cover document for a package I have this day sent to your address, by the Messagéries Maritimes. This package is addressed by Monsieur Pasteur, member of the Institute of France, and is destined for Messieurs Loir and Germont.' [234]

The box contained Pasteur anthrax vaccine, but the Clerk of Records didn't know this. Unable to ascertain, from the note, the contents or purpose of the box, he opened the second 'Confidential' letter that accompanied it. This letter discussed a number of matters in French, and the clerk gleaned that it contained instructions from Pasteur to Loir and Germont to use the contents of the box. Realising that he should not have ignored the 'Confidential' status of this letter and opened it, the clerk discarded the incriminating envelope that had borne it.

The Clerk of Records then carried the box, the note and the opened letter to the office of Chief Clerk Edward McKenny, to let him decide how to handle the matter. The second most senior administrator in the Chief Secretary's office after Principal Under Secretary Critchett Walker, Edward McKenny was a civil servant with thirty-five years' service. He promptly instructed the Clerk of Records to have the Post Office deliver the box to Loir and Germont at Albany Chambers in Bligh Street. But he neglected to include the letter addressed to them, instead deciding to consult his boss, Critchett Walker, about what he should do with the letter now that it had been opened.

Over the next few days, Post Office personnel attempted unsuccessfully to deliver the box into the hands of Loir and Germont at Albany Chambers. Loir and Germont were of course absent, and had been since 7 July, conducting their experiments on Rodd Island. Unable to locate them and make the delivery, the Post Office then returned the box to the Chief Secretary's Department.

At 5.00 on the evening of 18 July, eight days after the box had arrived in Sydney, an exasperated Chief Clerk McKenny personally took the box to Albany Chambers. He was aware that Frank Hinds was staying with Critchett Walker in the flat next door to that of Loir and Germont so, when he received no response at the Frenchmen's flat, McKenny rang the bell to Walker's residence. Hinds answered and McKenny handed the box over to him. McKenny also gave Hinds the letter from Pasteur addressed to Loir and Germont which was open and without its original envelope.

Before he departed, McKenny had Hinds sign for box and letter, and in turn gave him a note from Critchett Walker.[235] The note read:

'This letter was opened in this office. It has not been read or translated, and has not been seen by anyone except the Chief Clerk and myself. C.W.'[236] There, thought Edward McKenny, the matter would rest. He was to be proven spectacularly wrong.

By coincidence, that same day in Paris, Louis Pasteur realised that, while he had informed British Ambassador Lord Lytton in his 11 June letter that he had dispatched the box of anthrax vaccine to Sydney, he hadn't forewarned Adrien Loir that it was coming.

By this stage, Pasteur was almost totally preoccupied with his new institute. He was regularly travelling out to Grenelle to visit the growing complex, and there he would limp around the construction site with the aid of a walking stick; skullcap on his head, trying to visualise the finished structures and to picture a vibrant institution full of brilliant young people doing wonderful work for France and for mankind. And he would fret about costs and the still inadequate subscription fund. The news that Pasteur seemed assured to win the Australian rabbit prize had generated a new wave of donations in France, but the total was still well below the target set by Pasteur two years earlier, and his desperation for and obsession with the rabbit prize had not diminished.

It appeared to Pasteur that the institute buildings might well be completed. But what use was an empty shell? There would be little left in the fund once construction had been paid for. Where was the money to come from to cover equipment, salaries and other ongoing operating expenses if the Australian rabbit prize was denied him? There was still a possibility that the French Government would step in and wrest control of the institute from Pasteur's hands, in the name of national honour, to equip and staff it. For the Elysee Palace was preparing to make the planned November opening of the institute a State occasion.

With these worries burdening his mind, it was no wonder that when Pasteur had sent off the anthrax vaccine – which, in relation to the bigger rabbit prize picture, was entirely unimportant as far as Pasteur was concerned – he had forgotten to alert his nephew that the vaccine was on its way to Australia. On 18 July he corrected this oversight with a brief telegram to Sydney. As usual, to free Pasteur from any cost apart from that involved in a cable from Paris to London, the 18 July telegram would be transmitted to Loir and Germont via the New South Wales Agent-General in London and the office of Sir Henry Parkes in Sydney.

Ever since his insolvency the previous year, Sir Henry Parkes had been living at the inner-Sydney suburb of Balmain, not far from Iron Cove and Rodd Island. The seventy-two-year-old's home, Hampton Villa, was a leased Gothic revival mansion, the rent being paid by the benevolent fund his supporters had set up for him the previous November. Parkes shared the house with three adult daughters, who tolerated the frequent visits of their father's thirty-four-year-old mistress, Eleanor 'Nellie' Dixon, and her three young children.

On the chilly midwinter evening of Thursday 19 July, a delivery boy in the service of the New South Wales Electric Telegraph Office came to the Premier's door with a telegram. From London, the telegram was addressed to 'Eveniret'. In New South Wales Government circles, Eveniret was a code word signalling that the telegram had been sent from the Agent-General in London to the Premier in Sydney and was of top priority. The Telegraph Office had standing orders to always deliver Eveniret telegrams to the Premier no matter where he was at the time, and so occasionally they reached him at home out of business hours.

A servant brought the telegram to Sir Henry and, with a fire crackling in the grate, the Premier used a letter-opener to slice open the envelope. Unfolding the telegram, which was handwritten on a pink telegraph form, Sir Henry found that it was a message from Louis Pasteur to his representatives in Sydney. 'Steamer *Oceanien* takes box vaccine and letter addressed to Sir Henry Parkes,' it read. 'Acknowledge receipt immediately. Do not make demonstrations Cumberland disease until Rabbit question is terminated. Pasteur.'[237]

Parkes sat for a time, deep in thought and absently stroking his Irish setter Maori as the dog lay beside his master's chair. As a matter of course, every telegram that came in from Pasteur via the Agent-General in London for Pasteur's representatives was being read by Parkes and then copied by the Premier's staff before it was forwarded to Loir and Germont. Technically, this constituted spying but, as the telegrams were addressed to the Premier, the temptation to read them was too great. This process allowed Parkes to learn of Pasteur's instructions to his representatives even before Loir and Germont received them. As for the copying, Chief Clerk McKenny saw this as a way of keeping track of the cost of the Pasteur telegrams which his office had to bear – on the copy of this latest telegram, the clerk who transcribed it would write in red ink, '34 words, £12.6.6 (twelve pounds, six shillings and sixpence).'[238] But now Parkes took a step that went beyond simple surveillance of letters and telegrams.

Since Adrien Loir had proven that Cumberland disease was anthrax, Parkes had been eager for Pasteur's men to carry out the anthrax vaccine field trial that Frank Abigail had discussed with Loir. As Parkes was aware, Pasteur's anthrax vaccine had finally been handed over to the Pasteur Mission in Sydney just the day before. Pasteur's anthrax vaccine could be of incalculable help to New South Wales. And it would help Pasteur, who could expect to make a tidy profit from the sale of the vaccine right across this continent where sheep and cattle outnumbered the 3.5 million human inhabitants thirty to one.

But here in this telegram, Pasteur was instructing his team not to proceed with that anthrax trial until the rabbit competition had been settled – in his favour, he hoped. Wily Sir Henry Parkes knew that even if the troublesome Rabbit Commission did award Pasteur the prize – by no means a certainty – the competition rules meant that it could not be finalised until the winning remedy had been successfully tested for twelve months. And Loir and Germont had only commenced their experiments twelve days before. To Parkes' mind, there was no reason to delay the anthrax trial another day, let alone twelve months.

After due deliberation Parkes, the shrewd elder statesman of Australian politics, folded the telegram and deposited it on a shelf. There it would stay for the time being. The next day, the Premier would instruct Mines Secretary Abigail to tell his Stock Department to push forward with anthrax trials in the Riverina. As far as Parkes was concerned, by sitting on Pasteur's 18 July telegram he was doing everyone a favour, including Louis Pasteur.

17

THE RODD ISLAND EXPERIMENTS

OBLIVIOUS TO THE SABOTAGE and subterfuge being practised in various circles in Sydney, Loir and Germont were pressing on with their experiments on Rodd Island. In many ways the island proved to be an idyllic location for a research station. Considering that the facilities had been thrown up in less than two months, the station had a remarkably permanent look to it. A road had been blasted through the sandstone from the wooden jetty to the island's level summit. To save time, the buildings on the peak had been constructed with double walls of corrugated iron packed with sawdust. Inside, they were lined with felt. Pipes ran from every roof to water tanks which collected rainwater, the island's only source of fresh water.

The residency, the accommodation building, was on the western side of the island. Built on stilts and surrounded by a shading verandah two metres wide, this wing contained three bedrooms and a bathroom, a library/office, and a large dining room. 'It was all very comfortable,' Loir was to say.[239] The rectangular residency was connected by a covered walkway to a separate, smaller building containing a kitchen, a pantry and a cook's bedroom. This was occupied by Frederick Bell, the Irish cook hired by Camac Wilkinson, and his Irish wife. Mrs Bell was expected to act as housemaid as part of her husband's salary package of £1 and 15 shillings a week plus free accommodation and food.[240]

When Alfred Pearson had devised the preliminary rough design for the research station he had forgotten to allow for accommodation for the island's other staff member, a handyman. This man's job was to feed animals, burn carcasses, empty the lavatory pans, and generally keep the station in running order. So, while that handyman, one F. Ambrose, would eat with the cook and his wife, because of Pearson's oversight he had to sleep in a tent at the southwestern end of the island.

The laboratory building sat on the peak's southern side. Of similar design and construction to the residency, it contained a laboratory, a microscopic room fitted with a Zeiss microscope, a storeroom, a washhouse, and a furnace for the cremation of animal carcasses. The laboratory's scientific facilities consisted of tables, sink, incubator stands, and Bunsen burners. Gas was used in the laboratory's burners and to heat the culturing incubator; it also provided light and heat in all the buildings. After the failed attempt to pipe coal-gas to the island, gas cylinders were installed, but these kept running out at inconvenient moments. So eventually Oscar Katz installed a German apparatus, a Muller's Alpha Gas-Making Machine. It turned petroleum into gas – which the Bunsen burners would not burn well, requiring their replacement by Fletcher's burners.[241]

At the northern end of the island stood a stable building, also of corrugated iron, where the larger test animals were kept. Beside the stable there was an aviary – because the Rabbit Commission was determined to test chicken-cholera on birds even though birds were not included in the rabbit competition rules. The station's main buildings formed three sides of a quadrangle, and this quadrangle was covered on all sides by fine metal gauze. The enclosure's roof was also gauze, so that a compound was created; Adrien Loir nicknamed it 'the birdcage'.[242] The compound, in which a man could walk upright, was in effect a concentration camp for rabbits. The test rabbits were collected by the Rabbit Branch of the Lands Department in the country near Hay and transported in crates to Sydney by rail. Artificial concrete burrows were dug in the compound's sandy soil and covered with planks so they could be easily opened for inspection. Several small trees, gums and geebungs, had been left in the sandy compound to give it a semblance of a natural environment.[243]

To the east, south and west of the island, the mainland was only a matter of several hundred metres away. On the south and west houses were scattered along the mainland's shoreline. Parklands of the Callan Park Insane Asylum spread along a peninsular on the eastern side; some of the larger animals used in experiments on Rodd Island were kept there prior to transfer to the island. Despite the relative closeness of civilisation, Rodd Island possessed an atmosphere of exotic isolation. On its northern side were several secluded swimming places; fish could be caught from a small boat out on Iron Cove; and in the early morning, ducks and other water birds were in profusion all around the cove. For Adrien Loir, who had spent the last six years living in a tiny loft room at the ENS on Paris's Rue d'Ulm as the veritable prisoner of his uncle, the natural beauty and freedom of Rodd Island made it an unexpected paradise, even though it was winter.

In the wake of the Rabbit Commission ultimatum, Loir had been surprised and delighted by the unheralded advice from the Lands Department that the Pasteur Mission could commence Pasteur's experiments on Rodd Island on 7 July. He had no idea that he had Premier Parkes to thank for this sudden breakthrough. But it had allowed him to move into the experiments while keeping to himself the fact that Pasteur had been prepared to surrender and allow the Commission to carry out his experiments. Nor had Loir contacted Pasteur. As far as his uncle in Paris was concerned, Loir and his colleague Germont were now acting as mere spectators while the Commission conducted the experiments. Loir would let Pasteur know what he and Germont had done after the event, once they had completed their experiments on his behalf.

From that very first day on the island there was discord between the Pasteur team and the Rabbit Commission's man, Dr Oscar Katz. In quiet moments when nothing was happening, Katz would lock himself in the laboratory and conduct experiments of his own. Germont demanded that he and Loir be allowed in to see what

Katz was up to – both sides had agreed that each would be allowed to watch all chicken-cholera experiments conducted by the other. But Katz refused to admit the Frenchmen, declaring that what he was doing was none of their business.

Katz complained to the Experiment Committee that he was being harassed by Pasteur's representatives, prompting Camac Wilkinson to tell Loir and Germont that they could not assume the right 'of visiting the laboratory at any time in connection with the experiments of the Commission'. He 'pointed out that such indiscriminate visiting to experiments of this nature could not be permitted, but that stated times for visits must be arranged'.[244] Such an arrangement would allow Katz to hide what he was doing before Loir and Germont entered the laboratory. For, as only Wilkinson and the Experiment Committee knew, Katz was secretly conducting experiments in the Rodd Island laboratory with Robert Koch's chicken-cholera and rabbit septicaemia microbes from Germany. When Louis Pasteur learned of Katz's objectionable behaviour via Loir's next letter, he was to complain, 'How is it possible that in this Commission, not one member is revolted by the conduct of this German doctor?'[245]

Loir and Germont meanwhile worked through a range of chicken-cholera experiments prescribed by Pasteur – his first two original experiments and the added contagion experiment. Once those experiments had been completed they could move to the final stage – a field trial on an inland sheep property in an area of no more than 500 acres enclosed by rabbit-proof fencing and infested with rabbits.

Because their experiments on the island would be observed by Rabbit Commission representatives, Loir took a step back and continued to allow Germont to be seen as the front man. While Loir had become, via Pasteur's tutelage, one of the most experienced microbiologists in the world, technically he was merely a laboratory assistant while Germont was a qualified doctor of medicine. Letters and telegrams would always go out under Germont's name first, and it was Germont who took charge of the physical experiments with animals. But it was always Loir who carried out the more important laboratory procedures. And it was always Loir, Pasteur's nephew, who was making the decisions and calling the shots for the team behind the scenes.

As they worked, the two Frenchmen became accustomed to being constantly shadowed by gangly Oscar Katz. Notebook in hand, Katz watched their every move, writing down everything they did and when, and tabulating each result with Germanic precision. There is no record of Loir or Germont as much as conversing with Katz, so it's unlikely they even shared the same table with him in the residency's dining room. It was as though a state of war existed between the two sides.

Camac Wilkinson came out to the island whenever he could get away from parliament and the university. The other Experiment Committee members came too: Pearson, Bancroft, Quin, and most regularly of all, Fred Dillon Bell, who watched every experiment with the fascination of an amateur who desired to learn all he could about microbiology.

The first Pasteur experiments were designed to prove that chicken-cholera was an effective means of destroying rabbits, both via food treated with chicken-cholera broth and via contact with rabbits that were infected with chicken-cholera. Loir began at 3.00 pm on 7 July by feeding three rabbits with cabbage leaves laced with chicken-cholera broth. One rabbit died within thirty hours, one within thirty-seven hours, and the third after forty hours. Autopsies revealed they had died from chicken-cholera. Also on 7 July, Loir injected five rabbits with chicken-cholera then placed them in a wooden box one metre square and equipped with a wire mesh top. This box was placed on the paved floor of the stable. Twenty uninfected rabbits were put in the same box as the injected rabbits. All were subsequently offered untainted food and water, twice daily.

The first injected rabbit died within eight hours, and all five were dead within fourteen hours. Of the twenty rabbits that had not been injected nor fed poisoned food, five died over the next fifty-six hours. The implication was that these five had acquired chicken-cholera by contact with the sick or dead rabbits, but no autopsy was conducted to prove it. At the same time, a control experiment was run with twenty-five rabbits placed in a box identical to the first. Neither injected nor given poisoned food, these rabbits were fed and watered twice daily. After seven days, three of these rabbits had also unaccountably died, complicating the experiment.

The Experiment Committee would later declare that many rabbits collected at Hay for the Rodd Island experiments were already unhealthy and in the latter stages of starvation. A large number would die at the island, declining to eat the food offered them. Yet at one of the first meetings of the full Rabbit Commission, on 17 April, long before the Rodd Island experiments began, Queensland commissioner Dr Joseph Bancroft, an experienced naturalist, had warned: 'Wild rabbits will die in confinement, and they do not give us an opportunity for making experiments.'[246] Bancroft felt that experiments should be carried out with domesticated rabbits but, as there was a not a ready supply of the domesticated variety, to save time and trouble Bancroft's warning had been ignored, and would continue to be ignored, by the Experiment Committee.

On 17 July, Loir placed five rabbits that had been fed food infected with chicken-cholera in a metre-square box with twenty other rabbits. One of the poisoned rabbits was dead within twenty-seven hours. By 24 July, nine rabbits had died, the carcasses of all dead rabbits having been left in the box; a further two would be dead within another twenty-four hours.

On 24 July, microscopic examination of the blood of three of the deceased rabbits that had not been fed infected food proved they had died from chicken-cholera. On 24 July, two fresh rabbits were injected with blood from two of these last three dead rabbits. Both died within seventeen hours, and their blood showed the presence of chicken-cholera. Another fresh rabbit was then injected with the blood from one of these last two dead rabbits, and it was dead within twenty hours, again demonstrably from chicken-cholera.

By 27 July, when this experiment concluded, fourteen of the rabbits used in this experiment were still alive, including three control rabbits and one of those that had originally been fed food laced with chicken-cholera broth.

Starting on 17 July, Loir also ran another experiment concurrent with those focused on rabbits. This experiment was designed to prove that, as Pasteur had guaranteed over and again, chicken-cholera was not dangerous to domestic animals. A horse, a cow, a sheep, a goat, a pig and a dog were all given food laced with chicken-cholera broth. As a control, two rabbits were fed with food infected from the same culture. The control rabbits were dead within twenty-five hours and microscopic examination of their blood proved that chicken-cholera was the cause of death. All of the domestic animals were at that time in good health.

Six more control rabbits were immediately fed infected food. Three of these died overnight. On the 19th, the domestic animals were again given food to which chicken-cholera broth had been added, as were three more control rabbits. One of these rabbits died overnight and another rabbit injected with its blood also swiftly died. Control experiments of this type continued until the morning of 28 July. During the control experiments, twenty-six of thirty-six control rabbits died.

In the first week of August, all the surviving control rabbits would again be fed with infected food, and all eventually died, some after successive feedings. When at 9.30 am on 28 July, Loir and Germont wrapped up the domestic animal experiment, all the domestic animals under test were in good health, eating well and looking lively.

That morning, in pouring rain, Loir and Germont took a boat back to Circular Quay and then a cab to their rented Bligh Street flat, to catch up with their mail and for a taste of civilisation. They had to go to the Post Office for personal mail – each received such a quantity of mail from France that the clerks of the Post Office's *poste restante* section only had to see the Frenchmen's faces at the collection window to recognise them and hand over their bundles of letters without asking their names.[247] Apart from letters from Louis Pasteur, Loir was receiving mail from his aunt Marie Pasteur, from friends, and from his mother. Amelie Loir missed her only child so much that every time she received a letter from him in Australia – he wrote to her roughly once a fortnight – she would mark the date of receipt on her wall calendar.[248]

Official mail from the New South Wales Government was hand-delivered to Loir and Germont at Albany Chambers, and while they were at their flat on 28 July a letter written that same day by Alexander Bruce, Frank Abigail's Chief Inspector of Stock, was delivered to them. Alexander Bruce was a tall, dour Scotsman from Aberdeen, a deeply religious Congregationalist with a heart of gold. When hundreds of colonists died in a typhoid outbreak in 1889, Bruce and his wife would adopt three children orphaned by the epidemic and add them to their existing brood of twelve children. Nicknamed 'Tiger' by his friends and subordinates because he never let go once he got his teeth into something, Bruce lived for his work. When he wasn't in his office or travelling the colony on Stock Department business he was locked in his study at home poring over official documents or reading journal articles.[249]

Bruce had been informed by his superiors that the anthrax vaccine had arrived from Pasteur in Paris so, in his 28 July letter, Bruce asked when Pasteur's delegates might be ready to proceed with the anthrax vaccine trial, what facilities and stock they would require, and what program of experiments they would adopt. Loir and Germont had long before worked out their requirements so Germont immediately wrote back asking for forty sheep and ten head of cattle for the trial, and setting out the program they would follow – a program prescribed for them in Pasteur's delayed letter of 26 May that had accompanied the box of vaccine, and that was very similar to the program of the famous 1882 Pouilly-le-Fort trial.

'Tiger' Bruce passed Germont's reply on to his minister, Frank Abigail, who worked with characteristic speed. In the first week of August, while Loir and Germont were completing their last chicken-cholera experiments with the control rabbits on Rodd Island, Thomas W. Hammond of Junee, a justice of the peace and President of the Murrumbidgee Pastoral and Agricultural Association, was approached by Abigail to provide a site for the anthrax trial on his Riverina property. Hammond immediately agreed, setting aside thirty acres adjoining the railway line a kilometre from the town of Junee, which at the time was called Junee Junction.

Within another two weeks, Frank Abigail had put together a Cumberland Disease Board of five members to oversee the trial and report back to the government. Unlike the Rabbit Commission, this board was filled with men who would view Pasteur's science without bias or private agendas. Loir and Germont were already acquainted with four of the five board members – Chief Inspector of Stock Alexander 'Tiger' Bruce, Government Veterinarian Edward Stanley, Government Analyst William Hamlet, and Arthur Devlin of Uarah Station.

The fifth member of the board, and its chairman, was squatter John de Villiers Lamb, younger brother of Adrien Loir's new friend, Walter Lamb. Although he lived at Chatswood on Sydney's North Shore, John de Villiers Lamb had interests in a score of sheep and cattle properties in several colonies, land holdings that collectively covered thousands of square kilometres. A founding member of the Union Club, where he had probably already met Loir, Lamb was also a director of numerous companies whose interests ranged from insurance to gas, kerosene to banking. A tall, powerfully built man, Lamb rode and sailed competitively and was an officer in the New South Wales Volunteer Corps, in the colony's part-time army. Larger than life, J.D.V. Lamb was a man who would prove fascinating company for Adrien Loir.

Lamb was also Chairman of the Sydney Board of Sheep Directors, and several of his properties were affected by Cumberland disease. He was probably the squatter who said to Loir, 'Births cannot make up for the losses of Cumberland disease.' When Loir asked him what he meant by this, he replied, 'The mortality rate due to anthrax is on average between thirty and thirty-five percent on a station where I have some financial interest, and the births cannot compensate for the losses.' A number of squatters to whom Loir spoke quoted similar figures, which surprised Loir. In France, the mortality rate from anthrax prior to the introduction of vaccination had only been ten to twelve percent, so Loir set himself the task of finding out why it was so high in Australia.[250]

Even before the Governor of New South Wales officially appointed the Cumberland Disease Board on 14 August, the

Sydney Morning Herald of 9 August was enthusing: 'The proposed experiment, which is similar to one carried out many years ago on the Continent by M. Pasteur himself, is warmly approved by the Minister for Mines, Mr Abigail, who describes it as being of the greatest importance.'

Meanwhile, on 4 August, as Loir and Germont were wrapping up their experiments on Rodd Island, Dr Oscar Katz impatiently asked to be allowed to take a sample of heart blood from one of the rabbits that had recently died from chicken-cholera in the French team's experiments. Katz said that he wanted to proceed with the experiments that Camac Wilkinson and the Rabbit Commission's Experiment Committee planned to pursue. Pasteur had previously said that Loir could give the Commission this sample once *all* his own experiments had been completed. Although the first two stages of the Pasteur experiments were now at an end, there was still Pasteur's prescribed experiment in open country to be conducted. But with Katz hounding Loir for the sample, and with Camac Wilkinson warning him that if he did not hand it over the Commission would not give its permission for Pasteur's open country experiment, Loir agreed and the Austrian was provided with the heart blood sample.

The Commission had given its approval for the employment of an assistant to Chief Expert Officer Katz, at £16 per month plus room and board. As Loir and Germont left the island on 4 August, Katz was beginning the Commission's experiments with that assistant's help. The official Rabbit Commission reports never identified Katz's assistant - quite deliberately, because the appointment proved to be improper. Oscar Katz was to inadvertently reveal his assistant's identity the following year when he thanked him in writing for his services – it was Fred Dillon Bell, New Zealand's representative on the Rabbit Commission, Experiment Committee member, and enthusiastic amateur scientist.[251] This was equivalent to a member of a jury acting as assistant to the policeman in charge of a case on which the juryman would later sit in judgment. Katz and Bell – whose appointment was sanctioned by Experiment Committee Chairman Camac Wilkinson – took up residence on Rodd Island where they were to spend the next few months.

Through August, while Katz and his assistant were working on their experiments on the island, various buildings and rabbit-proof fencing were being thrown up at Junee on the thirty-acre paddock on Thomas Hammond's run, to create a temporary research station for Adrien Loir's anthrax vaccine trial. Carried out at Mines Department expense and supervised by Cumberland Disease Board member Arthur Devlin, the work was due to be completed in time for the trial to begin on 4 September. All through the month of August there was much talk about the trial in the press of all the colonies, and a number of squatters whose cattle were suffering from Cumberland disease indicated their interest in attending. The governments of Queensland, Victoria and Tasmania announced that they were sending representatives to join New South Wales Government delegates at Junee in September.

On Saturday 18 August, four days after Government House announced the appointment of the members of the Cumberland Disease Board, the Rabbit Commission officially sat again for the first time since June. The sitting took place in Melbourne and, apart from Camac Wilkinson and Secretary Hugh Mahon who both took the train down from Sydney, the only other members of the Commission present were Victoria's three representatives: Albert Pearson, Edward Lascelles and the Acting President, Professor Harry Allen. At this meeting, Dr Normand MacLaurin's written resignation was received, along with a note from Principal Under Secretary Critchett Walker advising that the New South Wales Government proposed to accept that resignation. Harry Allen's assumption of the role President of the Commission would soon be formally confirmed by a vote of the commissioners.

Also at this sitting, a still smarting Camac Wilkinson told his fellow commissioners how he had been called before Premier Parkes and rebuked for the treatment that Pasteur's delegates had received at the hands of the Commission. Worse, he said, was the way Parkes had gone over the Commission's head and allowed Pasteur's people to undertake their experiments on Rodd Island from 7 July, completely ignoring the Rabbit Commission's ultimatum. Secretary Hugh Mahon also told the meeting that the Lands Department – still controlled directly by Sir Henry Parkes – had also taken out of his hands the task of cataloguing the 1400 rabbit destruction devices proposed to the Commission as competition entries.

Professor Allen and his colleagues were highly indignant at hearing all this. How dare the Premier of New South Wales interfere with the workings of a royal commission involving all the colonies! The commissioners were also peeved that the Parkes Government was supporting the recently announced Pasteur anthrax vaccine trial. To anti-vaccinists this was an unwelcome harbinger of the widespread use of vaccination in Australia. To the rabbit commissioners, it was almost as if Parkes was thumbing his nose at them. In response, the commissioners unified behind Camac Wilkinson.

To signal the Commission's defiance of the New South Wales Premier, and its support for Camac Wilkinson, Albert Pearson, Wilkinson's cohort on the Experiment Committee, proposed that a copy of a previous Commission resolution 'be forwarded to Sir Henry Parkes in confirmation of the action of the Chairman of the Experiment Committee'.[252] This was a resolution declaring that Pasteur's representatives would not be authorised to carry out experiments in the open until laboratory experiments had proved that chicken-cholera would spread effectively from rabbit to rabbit.

Unanimously, the other commissioners present supported Pearson's motion. Via this resolution, the Commission was telling Parkes that they still had the upper hand, that they had the final say on the future of Pasteur's rabbit remedy, and not Parkes. The commissioners didn't stop there; they also gave a ringing endorsement to the commissioner berated by Parkes – Wilkinson. Previously, Dr MacLaurin and Secretary Mahon had been authorised to sign cheques on behalf of the Commission. In the wake of MacLaurin's resignation, the commissioners at this meeting voted that Camac Wilkinson now be authorised to sign cheques in MacLaurin's place.

Almost as an aside, Harry Allen also said that the memo that Wilkinson had sent Premier Parkes explaining the Commission's actions should be included in the Commission's report. When

Wilkinson oddly responded that he had sent his only copy of that memo to Pasteur's representatives – for what reason he did not make clear – Allen instructed Secretary Mahon to write to the Colonial Secretary's office seeking a copy of Wilkinson's memo. At the 16 October sitting of the Commission, Hugh Mahon would report that the Colonial Secretary's office had never received Wilkinson's memo. Mahon subsequently asked Pasteur's representatives for the copy, but they had never seen the memo either. Camac Wilkinson, it would seem, had never written or sent the memorandum, and had been caught out telling the sort of easily discovered lie that only a schoolboy would attempt to get away with.

But all this was months away when, at the 18 August Commission sitting, Professor Allen called on Wilkinson to read to his fellow commissioners Dr Katz's report on the experiments carried out on Rodd Island between 7 July and 4 August by Pasteur's representatives. Katz's report included appendices detailing each individual experiment, day by day, but the actual body of his report, on which hung the future of Louis Pasteur's rabbit competition entry and the £25,000 prize, amounted to less than 1400 words.

In his report, Katz criticised Pasteur's representatives for not using a German immersion objective apparatus and aniline dyes, which Katz said he had offered them, and for not carrying out autopsies on all dead rabbits, particularly in the early stages of the first experiment. Just the same, as far as proving that chicken-cholera microbes were fatal to rabbits when administered to them on their food, Katz wrote that he was 'perfectly satisfied' that Germont and Loir had achieved this objective.[253]

As for the second experiment – the contagion experiment which Pasteur added to prove that chicken-cholera was contagious among rabbits and would spread from rabbit to rabbit after the first rabbits had either been fed infected food or injected with chicken-cholera – this, Katz said, had been unsatisfactory in the way it had been carried out. He had two reasons for this belief - not all dead rabbits had been subjected to autopsy; and, at one point Germont had failed to mark one of five infected rabbits and so could not distinguish it from fresh, healthy rabbits, an error that had slightly compromised the experiment's results.

Even though a proportion of uninfected test rabbits had died as a result of contact with infected rabbits or from contact with their dead carcasses or their droppings, because of his complaints about the methodology, Katz reported that this experiment 'can, therefore, hardly be utilised for drawing any exact conclusions as to its practical value'.[254]

In the third experiment, which had set out to show that chicken-cholera was not harmful to domestic animals by testing chicken-cholera on a horse, a cow, a pig, a sheep and a dog, Katz reported that 'the result was a favourable one. At no time did any of the above animals show signs of illness; they presented always normal appearances'.[255] This experiment supported Pasteur's assertion that the disease was safe to use around domestic animals.

'I agree with the general tenor of Dr Katz's remarks,' said Camac Wilkinson after reading the report to his colleagues.[256] In other words, he did not enthusiastically endorse them. Katz's report had proven that in two experiments Pasteur had been right. The third experiment, whilst unsatisfactory in Katz's eyes, had not proven Pasteur wrong. This was not the ringing condemnation of Pasteur's method that Wilkinson, Pearson and Allen had been looking for.

That condemnation would now have to come from the Commission's experiments. With this in mind, Professor Allen now asked Wilkinson to report on the status of Oscar Katz's own experiments and the microbes that had been sought from Europe. Wilkinson replied, 'Cultures of microbes of chicken-cholera are in the possession of Dr Katz, and Dr Katz is now making experiments with them.' Wilkinson then corrected himself. The Commission's chicken-cholera experiments had not actually begun, he said, but were 'on the point of being carried out'. This was because 'the necessary cultivations have been in progress for some days'.[257]

In fact, Oscar Katz PhD had been struggling to cultivate chicken-cholera microbes from the heart blood of one of the rabbits that had died in Adrien Loir's experiments. He did succeed in the end, but it took him twelve days. From his later chicken-cholera paper printed in the Linnean Society's Proceedings for 1889, it would seem that Katz had never previously worked with chicken-cholera. Once he had perfected the culture, on 16 August without telling Wilkinson, Katz had actually undertaken his first experiment – a tentative test with a single hare, not a rabbit, to see whether his culture was sufficiently virulent, which it proved to be – the hare died on the 18th.[258]

In the Commission that same day, after Wilkinson's comment that Katz was still working on chicken-cholera culturing, discussion moved on to the competition submission that had come from Drs Ellis and Butcher. But Professor Allen's mind was still on the chicken-cholera microbes from Europe. What Wilkinson had just told the Commission did not quite add up. If Katz was in possession of cultivated chicken-cholera microbes one minute, but the next he was still striving to cultivate them, what was going on? After the Ellis–Butcher discussion, Allen asked Wilkinson to supply all Commission members with regular progress reports on all experiments, and then again inquired about the microbes which, so he had been informed, were being brought from Europe by Dr Carl Fischer for the Rabbit Commission's use.

Camac Wilkinson must have shifted uneasily in his chair before he answered. 'Dr Fischer recently arrived in Sydney,' he said, 'and handed to Dr Katz specimens which he said were the microbes of chicken-cholera and of rabbit septicaemia.'[259]

At no time did Wilkinson or any other rabbit commissioner name the original supplier of these microbes, both to protect the source and to protect themselves from any accusation from Louis Pasteur that they had gone behind his back and asked his greatest rival for microbes to use against him. It would be Oscar Katz himself, a year later, who would perhaps naively reveal: 'Dr Fischer, of Sydney, handed me on the 7th July, 1888, Agar-Agar cultures of these microbes, which he had brought from Koch's laboratory when in Berlin some time before.'[260]

From the outset, the plan of anti-vaccinists Camac Wilkinson and Professor Harry Allen had been to prevent Pasteur's representatives from testing their own microbes to prove Pasteur's rabbit eradication method. They had done everything in their power to make Adrien Loir hand over Pasteur's microbes – never realising how close they had come to achieving this goal when Pasteur had actually backed down – with the aim of conducting their own experiments to disqualify Pasteur's method. Back in May when Loir had refused to cooperate, Wilkinson had sought the alternative microbes from Robert Koch in Berlin, with Carl Fischer as the messenger. Once they had those, and with Pasteur's representatives shut out of the process, the Commission could have still carried out their own experiments, and as a result declared Pasteur's method a failure.

Two occurrences, both quite ironic, had upset Wilkinson's plan. Firstly on 7 July, the very same day that Katz had been handed Koch's microbes by Fischer, Pasteur's representatives had commenced their own experiments on Rodd Island as a result of Sir Henry Parkes' intervention. Wilkinson now revealed the second occurrence to the appalled commissioners at the 18 August meeting of the Rabbit Commission: 'On examination,' said Wilkinson of the microbes brought from Robert Koch in Germany, 'Dr Katz found that the cultures of chicken microbes were contaminated, and that those of rabbit septicaemia were dead. Hence, no experiments were made with such cultures.'[261]

Both batches of Koch microbes had proven useless. Oscar Katz had been forced to turn to the heart blood from a rabbit killed by the Pasteur microbes, and this was why Katz had hounded Adrien Loir to obtain the sample. Yet, could both samples from the famous Koch laboratory have really been faulty? It is possible that the rabbit septicaemia microbes were weakened after their journey across the world. But the Koch laboratory at the German Institute of Hygiene in Berlin was as hygienic as any anywhere. It seems strange that the chicken-cholera microbes contained in a sterilised, hermetically sealed jar, and kept in the care of Dr Fischer all the way to Australia, could have become contaminated by the time they reached Oscar Katz, when microbes that Adrien Loir had brought to Australia had survived.

There is, of course, another explanation. Several times over the next two years, Oscar Katz's skills as a microbiologist would come into question. Inadequate at best, incompetent at worst, there is a strong possibility that the Rabbit Commission's Chief Expert

Officer had himself contaminated the Koch microbes after they arrived as he fumbled to play the role of microbiologist. Had his critics within the Pasteur camp been aware of just how much Katz struggled to successfully complete his microbiological experiments for the Rabbit Commission, and later for another commission in Melbourne, they may well have suggested that Oscar Katz's PhD actually stood for 'Phony Doctor'.

Sir Henry Parkes, standing left, with his 1890 NSW cabinet. Daniel O'Connor, who was delegated to welcome Sarah Bernhardt to Sydney, is the other standing white-bearded minister, while the third standing figure is Sydney Smith, who took charge of the Rabbit Department in this ministry. Treasurer William McMillan is second seated figure from left. (Mitchell Library, State Library of New South Wales)

Louis Pasteur, Paris, 1884. (Courtesy Pasteur Museum, Paris)

Pasteur's nephew Adrien Loir, Brisbane, 1889–90. (John Oxley Library, State Library of Queensland)

Adrien Loir (right) conducting field trials with Pasteur's Australasian rabbit eradication method at the Pommery Champagne estate, Reims, France, Christmas 1887.
(*La Nature,* Paris, Spring 1892)

The Pasteur Institute, Paris, in 1889, shortly after it opened. Its opening was made possible by the Australian money obtained for Pasteur by his nephew Adrien Loir.
(Courtesy Pasteur Museum, Paris)

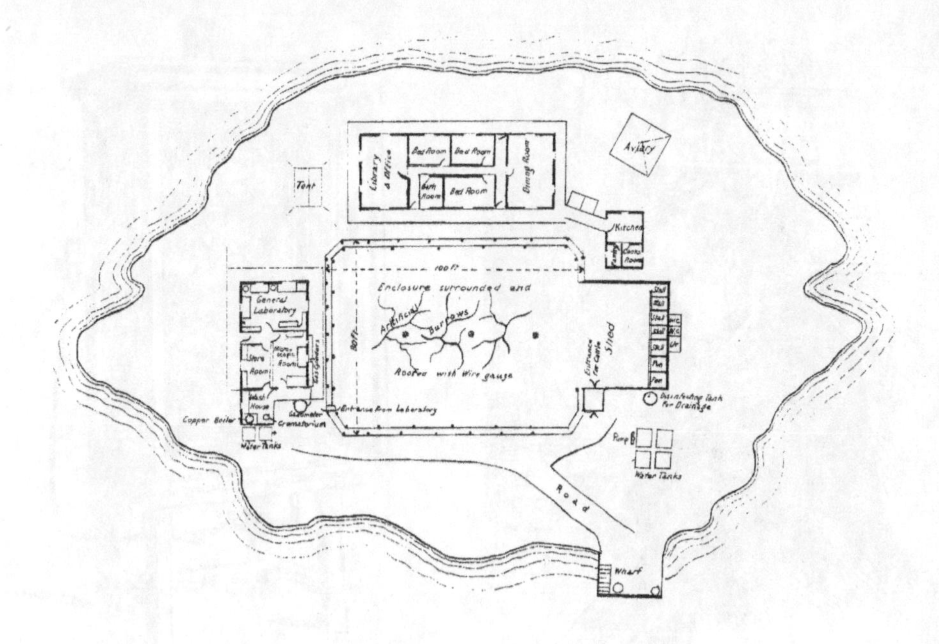

Ground plan for the Rodd Island Research Station, Sydney.

Rodd Island Research Station, 1888. Built for the Pasteur team's experiments.

Adrien Loir at work in the Rodd Island laboratory. (*Illustrated Sydney News*, 21 November 1891)

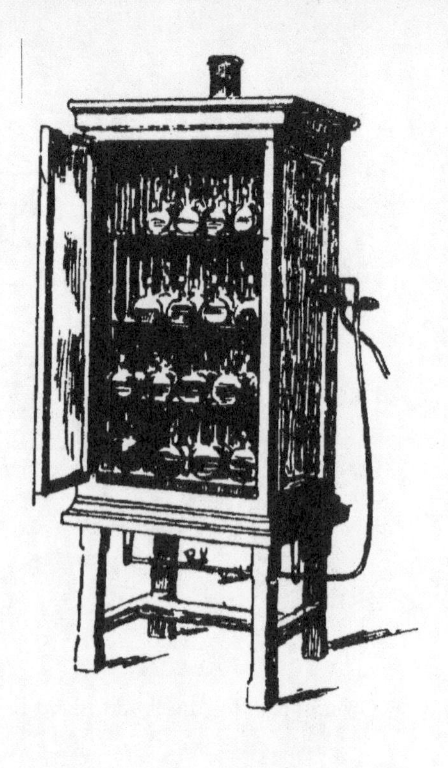

Loir's Microbe Incubator at the Rodd Island Research Station. (*Illustrated Sydney News*, 21 November 1891)

Kitchen and Residency buildings, Rodd Island, Sydney. (Louise Dando-Collins)

1889 cartoon from Sydney press lampooning the Pasteur rabbit eradication method. It is believed to depict the Rabbit Commission's controversial Chief Expert Officer, Dr Oscar Katz. (Adrien Loir scrapbook, Basser Library, Australian Academy of Science)

THE RABBIT CONFERENCE.

"It was resolved that in the opinion of this Conference the usual vegetable diet provided by Nature and the agricultural classes of these colonies is good enough for us, and that we vigorously abstain from trying any innovations in the shape of so-called Microbes that may hereafter be imported for our use."—Carried.

Newspaper cartoon on the Rabbit Commission, 1889. (Adrien Loir scrapbook, Basser Library, Australian Academy of Science)

John McGarvie-Smith in his laboratory, circa 1898.
(Courtesy McGarvie-Smith Institute)

18

SHOWDOWN WITH SIR HENRY

AFTER COMPLETING THE FIRST series of chicken-cholera experiments on Rodd Island on 4 August, Loir and Germont had returned to Albany Chambers in Bligh Street where they prepared to set off at the beginning of September for the anthrax vaccine trial at Junee.

Pasteur's delegates felt that they'd completed all they'd set out to achieve with the experiments on the island. As far as Loir was concerned, 'The delegates proved in the presence of the Commission that, in accordance with the conditions described in the competition and Monsieur Pasteur's statements, chicken-cholera could easily kill rabbits, whereas it spared horses, sheep, goats, pigs and cattle.' Loir felt that the outcome of the demonstration on the island had been beyond dispute. 'Following that demonstration,' he said, 'the Mission asked to be allowed to carry out the tests on the larger scale in the country.'[262] Camac Wilkinson had told him the Commission would be in touch in due course regarding that large-scale experiment.

On 9 August, back at Albany Chambers, Loir and Germont received a telegram from Pasteur, sent from Paris the previous day. Because Loir had not responded to his uncle's 25 June telegram – in which Pasteur had instructed him to surrender to the Commission's ultimatum and hand over their microbes – in this latest communication, addressed to the Agent-General in London, Pasteur asked that the contents of his message be passed onto both Germont and the Rabbit Commission. 'I accept the conditions in the letter of the Commission of 21 June,' he cabled. He was referring here to the Commission's ultimatum. He said, regarding Germont and Loir, 'I authorise them to allow the experiments prescribed by the Commission to be conducted and to furnish rabbits infected with chicken-cholera for experiments under the direction of the Commission.'[263]

Pasteur's instructions in this latest telegram had been superseded by events – by this time Loir and Germont had completed the Pasteur experiments on Rodd Island, had handed over a rabbit killed by chicken-cholera, and were awaiting Rabbit Commission permission to proceed with the open country chicken-cholera experiment on 500 acres. While a copy of this telegram had indeed been sent to Germont and Loir, the Colonial Secretary's office did not send a copy to the Rabbit Commission as Pasteur had requested – Premier Parkes did not want the Commission to know that Pasteur had buckled to their will and so weakened his position, and that of Parkes.

There can be no doubt that it was Parkes who personally decided who would receive this telegram: on the file, along with the copies of all the telegrams of this period, is Parkes' handwritten note: 'Refer to Dr Germont. H.P.'[264] A clerk's notation on the file for this particular telegram states, 'Copy sent to Germont, 9 August.' There is no reference to Parkes ordering that a copy be sent to the Commission. Likewise, the Commission's official report makes no mention of receipt of this telegram – all such communications were recorded – and it was never discussed in Commission meetings. Through Parkes' intervention, the Commission remained totally ignorant of the fact that Pasteur had been prepared to give in to their demands.

Germont did not reply to this 8 August telegram from Pasteur; he and Loir were waiting for Commission approval to move to the open country experiment after which they planned to let Pasteur know the good news. Instead, Germont and Loir sat down on the 9th and composed a letter to Pasteur in which they explained at length all that they had done in relation to both the chicken-cholera and anthrax experiments. They posted the letter that same day, hoping that Pasteur would forgive them for consistently disobeying his orders.

Hearing nothing back from Sydney and fearing the worst, Pasteur panicked. On 14 August, he cabled Germont: 'Execute instructions telegram eighth relating chicken-cholera. I propose by present telegram to Sydney Government another microbe also rabbit destructor but inoffensive birds.' Still desperate to win the prize, Pasteur would do anything to please the Rabbit Commission. He concluded his garbled cable with: 'This microbe is the one discovered by me in 1881.'[265] This was the only reference ever made to a variant of chicken-cholera that did not affect birds; Pasteur would never again mention it.

Still Germont and Loir remained silent. No reply to Pasteur's 14 August telegram was sent. Loir knew that it would be impossible to give Pasteur an accurate picture in a telegram, and was confident that once Pasteur read the letter of 9 August he would understand everything and, hopefully, approve of all that Loir had done, and forgive him for his disobedience.

On 23 August, a government messenger knocked on the Frenchmen's door at Albany Chambers and handed over yet another telegram from Pasteur. But this was the telegram of 18 July, the one that Premier Parkes had withheld. Only now, after Loir and Germont had committed themselves to conduct the anthrax trial, and with preparations at Junee almost complete, did Parkes release the telegram, thirty-six days after it had been received in Sydney. On its reverse corner there was a note in Parkes' handwriting: 'Mislaid. H.P.'[266]

François Germont did not have Adrien Loir's placid temperament. Tall, broad-shouldered and sometime boisterous, Germont had a temper, and now he lost it. For months, the Rabbit Commission had been working against the Pasteur team. Then there had been the debacle over the delayed delivery of the vaccine box, and the letter from Pasteur that had been opened in the Premier's office. And now this obvious withholding of Pasteur's instructions so that the Frenchmen would in their ignorance proceed with the anthrax trial. Germont wanted an explanation for this latest act of sabotage, and he wanted it now! He immediately stormed out the door and set off for the Premier's office. Loir, who could not physically stop the big man, had no choice but to join him.

When the two Frenchmen arrived at the imposing sandstone edifice that was the Colonial Secretary's Department, the burly Germont loudly demanded an immediate audience with the Premier. Parkes had the pair shown into his third-level office. Parkes' mahogany partner's desk sat in the middle of the large office like a ship in the middle of an ocean. It was covered with a thick scattering of folders and papers. Tall bookcases against the wall behind him were crammed with leather-bound volumes. More books were stacked on the floor, and piles of paper spilled onto the red carpet in front of the grey marble fireplace. Sir Henry, sitting beneath a five-pendant electric chandelier, asked what he could do for the Frenchmen. Germont, speaking rapidly and in heavily accented English, gushed out his complaint that the New South Wales Government was engaged in spying and sabotage.

Parkes made no attempt to respond to the accusation. 'I knew so little about it,' Parkes was later to say, 'that I refused to go into the matter except in the presence of the Under Secretary, and I at once sent for him to explain how these letters had been opened.' Critchett Walker's office was located just across the corridor from that of the Premier, and Walker soon arrived on the scene. "If there is one man stamped with the very image of personal honour,' Parkes went on, 'it is Mr Critchett Walker, and Mr Walker explained to these gentlemen in my presence, and at my request, all he knew about it.'[267]

Walker told Germont and Loir about the mix-up with the vaccine box and the two accompanying letters. He told them about the Clerk of Records' initial error, the problem with locating Loir and Germont, and Edward McKenny's attempt to set everything right. As for the delayed telegram – the straw that had broken the camel's back as far as Germont was concerned – that, said Critchett Walker, had simply been mislaid. 'Though they seemed a little huffed at first,' Parkes would later say of Germont and Loir, 'they went away leaving me under the impression that they were satisfied.'[268]

On the contrary, Loir was not at all satisfied. 'All that we demanded was a simple acknowledgment of the facts,' he was to say.[269] Yet it was naive to expect Parkes or his deputy to acknowledge that there had been foul play on the part of the government, which had without question been the case with the 18 July telegram. Back at Albany Chambers Germont wanted to take it further, but Loir, always the diplomat, talked him out of it. Yet the matter did not rest there; soon, two further incidents would raise Germont's ire.

Most of the letters from Pasteur to Loir and Germont reached them without further problems, but over the next six weeks two letters, one addressed to Loir and another addressed to Loir and Germont, in care of Sydney Post Office, would be clumsily opened and resealed before they were delivered. Both had the word 'Government' written on the front in pencil, in a hand other than Pasteur's. Sir Henry Parkes would later explain that the word 'Government' had been written on the envelopes by the postal authorities, who had sent them to the Colonial Secretary's office by mistake. The Clerk of Records had opened them, and once he had realised his error he had resealed them and had them delivered to Albany Chambers.[270]

Loir decided not to make an issue of these latest annoyances. 'The reception with which we met at the Colonial Secretary's office, when we made inquiries about the first letter, induced us to think another attempt was not worth the while', he would say six months later.[271] Loir, only interested in making this mission to Australia a success for his uncle, wanted to steer a safe course. Germont, on the other hand, was incensed by what he saw as the injustice of their treatment, and could not let this letter-tampering go without causing a stir. In the middle of September once the two resealed letters had been delivered, hot-headed Germont approached one of the Premier's political opponents, Legislative Assembly member and former Protectionist minister, William Lyne.

Lyne had been president of the colony's ineffectual Water Conservation Royal Commission of 1884–86, and Germont may have first become acquainted with him at a Sydney club when the subject of royal commissions came up – Germont not being an admirer of the current royal commission. Unaware that Sir Henry Parkes was probably Pasteur's greatest friend in the colony, Germont complained bitterly to Lyne about the Premier's interception of the letters from Pasteur, and urged the politician to attack Parkes in parliament.

On 20 September, William Lyne stood up in the Legislative Assembly and directed a question to Sir Henry Parkes: 'Is the statement true which has been made by Dr Germont to Mr Lyne, viz, that certain confidential letters addressed to Drs Germont and Loir, through the Colonial Secretary's office, were opened? If so, were they opened inadvertently, and by whom?'

Premier Parkes replied, 'Yes. The letters were inadvertently cut open with a great number of others addressed to the Department by the Clerk of Records, who usually opens the letters in the absence of the Principal Under Secretary. As soon as the mistake was discovered the letters were immediately sealed, and a certificate given to the effect that the contents had not been read or translated.'[272]

Lyne's question had been far from hard-hitting, and invited easy dismissal. More importantly Lyne, who was known to be slow on the uptake, had missed the Premier's real act of sabotage with the withheld 18 July telegram. The press did not pick up on Lyne's question, and the MP failed to ask another parliamentary question on the subject. The matter seemed to die. And Adrien Loir was happy to let it die. While Germont would continue to stew about foul play at the Colonial Secretary's office, Loir could see that complaining about Sir Henry Parkes and his government would not help Pasteur win the rabbit prize. And then there was the matter of the anthrax vaccine trial and the government contract for vaccine supply that Loir hoped would follow. Germont may have been an excellent microbiologist but he made a dreadful diplomat and an even worse businessman.

With delivery of the overdue telegram, even though it was thirty-six days late, Loir was presented with a problem. With just days to go before the anthrax field trial was due to begin, here Loir now had instructions from Pasteur, in this telegram, quite implicitly telling him not to undertake any such trial until the rabbit competition had been settled. Loir was a very bright young man. He knew there was a strong possibility the Rabbit Commission would rule against Pasteur's rabbit eradication method. Yet the possibility that he could secure huge anthrax vaccine contracts with the colonial governments and thereby earn hundreds of thousands of francs for the new Pasteur Institute was also extremely high. What should Loir do?

For the third time since his arrival in New South Wales, Loir decided to disobey his uncle. He would defy Pasteur's express instruction not to conduct the anthrax trial until the rabbit question was settled. And once again he acted this way because he had Pasteur's best interests at heart. Loir was to rationalise his decision by saying that the anthrax experiments were already too far advanced for him to back out. First there had been the experiments at the Government Laboratory, then the tests at Uarah Station. The Junee trial would be the last stage of a program he had commenced in May. Pasteur's telegram of 18 July had been received much too late, Loir would say the following year: 'It was handed to us only when in our ignorance of Mr Pasteur's instructions we had begun these experiments.'[273] So the Australian anthrax field trial would go ahead after all. Against Pasteur's wishes.

19

JOY AT JUNEE

FEELING THAT THEY MIGHT at last be getting somewhere, Loir and Germont left Albany Chambers on 3 September and took a train to Junee with the Sydney members of the Cumberland Disease Board. Frank Hinds did not accompany them. He was no longer a member of the Pasteur Mission to Australia. From later events, it is almost certain that in the confrontation in Sir Henry Parkes' office on 23 August, Germont had said something inflammatory about the relationship between Critchett Walker and his houseguest Hinds. And Sir Henry, to prevent a personal scandal that could ruin Walker's career, and a political scandal that would embroil the Premier, had immediately told Walker to cut his ties with the young English doctor. By the end of the following week, Frank Hinds had moved out of Albany Chambers, boarded an Orient Line steamer, and sailed away. Hinds, who had been nothing but a dead weight to Loir and Germont, went home to England and was never heard of again.

The train carried Loir, Germont and Pasteur's anthrax vaccine south to Goulburn. There, the western line swung right toward the vast dry interior of the continent. From Gouburn, Loir and Germont steamed west along shining rails that extended into the distant Riverina, via Gunning, Yass and Cootamundra. Seven hours and more than 300 kilometres after leaving Sydney, they reached the temporary research station beside the railway line, a kilometre east of Junee. Just a little above the east-west-flowing Murrumbidgee River, the country around them was flat but fertile, dotted with tall native trees – red gum, grey box and yellow jack. Here, Loir and his companions found newly built single-storey timber buildings equipped with bunks, a kitchen, and a hut that would serve as a laboratory.

The fifth Cumberland Disease Board member, Arthur Devlin, met the Sydney party here at the trial site. Devlin's property, Uarah, was another forty kilometres to the west. Tom Hammond, owner of the run where the trial was taking place, was here too, along with some of his station hands. All the people who had expressed a desire to witness the anthrax trial had been told not to bother turning up for the preliminaries – they should wait for the last stage, which would take place in a month's time. For the moment, it was just the board members, Hammond and his jackeroos who watched as Germont and Loir set to work on the morning of 4 September. Because this trial was going to be so public, Loir was again taking a back seat. Germont, a qualified doctor, would be seen to carry out the inoculations, with Loir acting as his assistant.

For use in the trial, the New South Wales Stock Department had purchased a number of sheep and cattle at Cootamundra, further east, in an area free of Cumberland disease. Early in the morning, before the heat of the day, Germont and Loir prepared to inoculate a number of these animals with Pasteur's anthrax vaccine. Germont, who had appropriated a broad-brimmed Australian bushman's hat, now stripped off his jacket and vest and went to work. Loir wore suit, vest, collar and tie, and the same jaunty, narrow-brimmed cap that he had worn at Madame Pommery's when he'd conducted his chicken-cholera experiment at Christmas: how long ago that seemed! The members of the board were also in suits and ties. The jackeroos, some of whom were ebony-skinned Aborigines, were in collarless shirts and grimy trousers held up by braces, and wore battered bush hats that had seen plenty of Riverina summers.

While this experimental trial procedure was new to the Australians, and to a certain extent to Germont, it was old hat to Loir; he had been working with the anthrax vaccine for years. Via trial and error, over the past seven years since Pouilly-le-Fort, the anthrax vaccination process in France had been perfected and turned into a smoothly run business. The vaccine that Pasteur had sent out to Loir and Germont came in two doses, the first weak, the second strong. Produced at Emile Duclaux's facility in the Rue Vauquelin, the vaccine supplies were regularly delivered, in pairs of glass tubes, to a private company in Paris, the Compagnie de Vulgarisation de Vaccin Charbonneaux Pasteur, in the Rue Lafitte, which handled sale and distribution of the vaccine to several hundred veterinary surgeons across France.

To build up an animal's immunity, French veterinarians began by injecting stock with the first, weak dose of the anthrax vaccine. Twelve to fifteen days later, they would inoculate the same animals with the second, stronger dose. This two-dose treatment had been devised by Pasteur because he believed that many animals would not have survived a single, powerful inoculation. With a success rate of ninety-nine percent in France, this two-dose vaccination gave animals immunity for between one and three years, and the number of sheep and cattle annually inoculated with Pasteur's anthrax vaccine in France had remained steady over the years – 305,000 in 1887.[274]

At the Junee trial on 4 September, Germont and Loir planned to inject twenty sheep and four head of cattle with the first, weak dose of the vaccine. Loir sterilised a Pravaz syringe by boiling it in a bucket of water over a cooking fire for ten minutes before filling the syringe with the clear vaccine from the glass tube marked '1st Vaccine'– each tube contained enough serum to vaccinate between one hundred and two hundred sheep.

The fencing surrounding the Hammond paddock at the test site was of post-and-rail construction. At a spot beneath the shade of a clump of red gums, one top rail was removed and the participants gathered around. A brawny jackeroo would lift a sheep and sit it on its rump on the second rail. Germont would inject the sheep on the inside of the right thigh, after which it would be lifted away to make room for the next ewe. The four cattle were then inoculated, while standing – behind the shoulder.

The test animals were watched closely over the next few days. Apart from temporary rises in temperature, all remained healthy. On 7 September, Loir and Germont returned to Sydney, where they remained for close to two weeks

By 18 September, most of the parties involved were back at the Junee test site for the next stage of the trial. Germont and Loir now gave the animals they had previously inoculated their second, stronger shots using the contents of the '2nd Vaccine' tubes. This time the injections went into the ewes' left thigh. On this occasion too, two sheep that had not been inoculated were injected with raw anthrax bacilli that Loir had harvested from the sheep blood collected at Arthur Devlin's Uarah Station in May, and that Government Veterinarian Stanley had been keeping in cool storage ever since.

These latter two sheep were dead by 8.00 pm the following day proving, as Loir had hoped, that the Uarah anthrax bacilli had lost none of their virulence. Government Analyst William Hamlet then drew blood, which now contained the anthrax bacilli, from these two dead sheep. Once again everyone went their separate ways. Back in Sydney, Loir prepared an anthrax cultivation from the sheep blood that Hamlet had taken at Junee. This cultivation, which would be used in the final stage of the trial, then went into Hamlet's safe-keeping at the Government Laboratory.

Waiting for Loir and Germont at Albany Chambers on their return from Junee was a new telegram from Pasteur. Dated 18 September, it read: 'My telegram ninth August was in answer to letters in English enclosed in Adrien's letter of 21st June. My following telegram was to fix date for new disease. Do not consider it otherwise. Have received your letter ninth August. Approve entirely what you have done.'[275] What relief this message must have brought Loir and Germont, but particularly Loir – his uncle had retained his faith in him. Pasteur had received their 9 August letter in which the pair had explained all that they had done and why, including disobeying his instructions. And Pasteur entirely approved of everything! What was more, it sounded as though, inspired by his nephew's defiance, Pasteur had regained his fighting spirit. On the last Saturday in September, Loir returned to the Junee experimental station with the anthrax culture as well as renewed confidence.

At 10.30 am on 30 September, observed by Tiger Bruce and Arthur Devlin, Germont and Loir injected three new sheep with anthrax bacilli. Loir, knowing from experience roughly how long it would take for the anthrax to kill, used varying quantities of the bacillus with each sheep so that at least one of the animals would die on 2 October, the appointed day of the trial's main demonstration.

On Monday 1 October, men began to arrive at the site by train, by carriage and on horseback. Tents sprang up all around the perimeter of the experimental enclosure to accommodate those who found the hotels of Junee full. Colonials kept arriving all through Monday. When the sun rose the following day, Tuesday 2 October, there would be more than two hundred interested observers waiting expectantly to see the final results of the Pasteur anthrax trial.

That same day in Paris, Louis Pasteur, totally unaware that Loir and Germont were on the verge of repeating the Pouilly-le-Fort anthrax trial in New South Wales, dictated a letter that was to be sent to Australia. The recipient was not Loir or Germont, but Alexander 'Tiger' Bruce, New South Wales' Chief Inspector of Stock. Bruce had quickly grown fond of young Adrien Loir since first making his acquaintance in May, and it seems that in August he had written to Pasteur to tell him what a great job his nephew was doing in Australia. Pasteur now sent a long letter in response: 'My representatives have only one thing to do, and I have sent them to do it, at the request of the government,' Pasteur told Bruce in this letter. That one thing, he said, was to demonstrate the efficacy of his chicken-cholera method against rabbits.[276]

'How to arrange for the manufacture of the fatal ingredient, how to mix and use at a distance those ingredient with all its properties, that is my secret,' Pasteur went on, 'about which the Commission is to see and know nothing for the present, and which I will only make known if the prize proposed on the 31st August, 1887, is awarded to me.'[277] Upon receiving this letter in November, Bruce would be so impressed that the great man Pasteur had written to him personally that he would send the letter to the *Sydney Morning Herald*, which published it. Ultimately, this letter was to do Pasteur more harm than good.

As Pasteur was composing his letter, at the Hammond property outside Junee Loir and Germont were being introduced to each of the men who'd come to watch the anthrax trial that Pasteur knew nothing about. They had travelled from four colonies – the five members of New South Wales' Cumberland Disease Board; a doctor from the Board of Health in Sydney; the Chief Inspectors of Stock from Victoria, Queensland and Tasmania, the latter being Rabbit Commission member Thomas Tabart; the Government Veterinarians for Victoria and Tasmania; numerous government stock inspectors; delegates from pasture and stock protection boards; squatters; and newspapermen. A carnival atmosphere prevailed. For Loir and Germont, it was time to put on a show.

That show began at 6.00 am on the morning of 2 October when one of the sheep injected with anthrax on the previous Sunday died. The second died at 3.30 pm, the third would die the following day. A little after 3.30, an autopsy was conducted on the second sheep, and blood taken. After the board members agreed that the microscopic evidence proved that sheep number two had died from Cumberland disease, thirty-nine sheep were herded into two pens around which the spectators crowded. All the sheep had numbered ear tags and the treatment of each individual animal was carefully recorded. Twenty of these sheep, so the spectators were told by a board spokesperson, had been inoculated with Pasteur's anthrax vaccine on 4 September, while the remaining nineteen were healthy sheep that had not been inoculated.

Starting at 4.45 pm, Germont and Loir injected all thirty-nine sheep with blood from dead sheep number two, injecting first a vaccinated sheep and then an unvaccinated sheep, and so on. All the sheep were then put in one pen. Six head of cattle were also injected with the same blood – the four that had previously been vaccinated and two unvaccinated beasts. Now it was a matter of waiting – the unvaccinated sheep, Loir and Germont confidently announced, would start to die the following day.

First thing next morning, everyone reassembled and eagerly observed the animals for signs of illness. All through the day they waited, without result, but the Frenchmen continued to exude confidence. The sun set. Then, at 8.30 pm, after dark and in the light of paraffin lamps, the first of the unvaccinated sheep was pronounced dead. All nineteen of the unvaccinated sheep died over the next two days. On the Thursday, with fifteen of the sheep dead and their vaccinated sisters walking over the carcasses without a sign of ill health, Tiger Bruce hurried into Junee and sent a telegram to his minister. The next day, a train came chuffing down the track from Sydney and pulled up right beside the experimental station. Smiling broadly, Mines and Agriculture Minister Frank Abigail stepped down from the train and, shaking hands with all and sundry, joined the audience for the final stage of the trial.

By the morning of Sunday 7 October, all the unvaccinated sheep were dead. One of the unvaccinated cattle was dead, and all the others were looking decidedly ill. At the same time, all the vaccinated sheep and vaccinated cattle were totally unaffected and in remarkably good health. Autopsies would be carried out by the board on all the dead animals, and it would be proved that every one had died from anthrax. In addition, at the request of one of the squatters present, one of the sheep that had been vaccinated in September was slaughtered and its blood tested – the test showed that there was no sign of anthrax microbes in the blood as a result of its vaccination.

The board declared the trial a complete success, and the gathered spectators cheered and applauded Germont and Loir. Minister Abigail, beaming from ear to ear, warmly shook the Frenchmen by the hand but probably avoided kissing them on the cheek; this was Australia, after all. 'He could not help showing his great satisfaction,' Loir was to write of Abigail's reaction, 'after all that had been said to denigrate Pasteur's experiments.'[278]

Abigail now convened a meeting in nearby Junee, in a hall beside the town's ornate railway station, with everyone cheerfully trooping into town to take part. Abigail, taking the chair, invited Government Veterinarian Stanley to address the crowd, and Stanley called on all present to agree to accept an anthrax vaccination program proposed by Loir and Germont. But Stanley added, that program could only go forward if every squatter present brought along a good number of stock to be vaccinated. He then asked how many sheep members of the audience would be prepared to bring to the program.

'During the following ten minutes,' Adrien Loir later wrote, 'I witnessed one of the most moving scenes in my career.'[279] Hands were raised and squatters competed with each other in their enthusiasm, as though this were an auction for a rare treasure.

'Three hundred sheep.'

'Six thousand sheep.'

'Twenty thousand.'

'When all were counted,' Loir wrote, 'altogether 260,000 sheep had been offered for vaccination.'[280] In the end it was agreed that perhaps half that number, 130,000, would be a realistic quota with which to launch the program, considering the work involved in producing the vaccine, in training vaccinators, and in carrying out the vaccinations in the field. Subsequently, over the period of a year, perhaps another 100,000 or more could be accommodated.

The two hundred Australians who attended the Junee anthrax trial would never forget those days in early October 1888. The story of the Pasteur trial would pass down through their families, with the roles of some of the participants inflating with time; in the version of the story told by some of Arthur Devlin's descendants, it would be said that Devlin had personally brought Pasteur's assistants out from France to conduct the trial.[281]

In the second week of October, feeling euphoric, Loir and Germont rode the train back to Sydney. Not only had the anthrax trial been a spectacular success, during their time at the Junee experimental site the pair had made many new friends and admirers. All five members of the Cumberland Disease Board would become Loir's good friends. He would grow especially close to Tiger Bruce and Arthur Devlin. This episode had been so completely opposite to the way the Pasteur Mission had been treated by the Rabbit Commission, it was as though the Frenchmen were in a different country.

Once they were back in the city, the pair anxiously awaited the Cumberland Disease Board's official report. They did not have to wait long. On 12 October, the report was released, concluding: 'After carefully watching the whole series of experiments and giving the subject the fullest consideration, the Board are unanimously of the opinion that Dr Germont and M. Loir have conclusively demonstrated the efficacy of M. Pasteur's "Vaccine of Anthrax" as a preventative against that disease, and therefore recommend its adoption and use.'[282] Loir and Germont joyfully cabled Pasteur with the news that he had at last won a battle in Australia.

The board's report had the effect of giving Loir and Germont an official seal of approval. The press immediately published the report's positive findings and described the beneficial implications for the colony of their seemingly miraculous demonstration at Junee. Virtually overnight, after six months in the Antipodes, Loir and Germont had gone from curiosities to celebrities. Government House invited Loir to dinner with His Excellency the Governor in October, and to a garden party in November. Private invitations flooded in from leading society people who wanted to meet these young French scientists who'd become glamorous identities.

Yet days passed without a word from Pasteur. Loir knew that his uncle must be busy with the final arrangements for the opening of the Pasteur Institute, which was due to take place on 14 November, but his silence after receiving such good news would have puzzled Loir. On 20 October, eight days after they'd cabled Pasteur, Loir and Germont finally received a telegram in reply to theirs, sent by Pasteur the previous day: 'Pasteur to Germont via new institute bearing my name. I can ensure now and in the future the supply of cultures for the anthrax vaccine for a fixed fee of one hundred thousand pounds. Wait for payment of that sum by the government before proceeding to the first practical experimentation on the hundred thousand sheep. Point out that the cultures will be produced in Australia by one or several young scientists trained by Pasteur Institute. All expenses to be met by the government.'[283]

As was his surreptitious practice, Sir Henry Parkes would have read this astonishing telegram before it reached Germont and Loir.

His reaction can only be guessed at. Was it laughter? Horror? Maybe both. Perhaps he thought that Pasteur had gone mad. To provide his anthrax vaccine to New South Wales farmers, the French scientist was asking for an upfront fee of £100,000 ($40 million in today's money), four times the rabbit competition prize, and roughly 2.5 million French francs. The total raised for the Pasteur Institute subscription fund from all sources over the two and a half years since the fund's inception had not even approached that amount. Pasteur's demand was outrageous.

Loir and Germont had just proven that the Pasteur anthrax vaccine would counter the dreaded Cumberland disease, and had built up the hopes of colonial farmers after it had been announced in the press that 130,000 sheep would be vaccinated immediately. Many people in Sir Henry Parkes' position would have seen Pasteur's telegram as a greedy attempt to blackmail the government and farmers of New South Wales. After receiving this telegram, many in Sir Henry's position would have been sorely tempted to wipe their hands of Louis Pasteur. And this was precisely what Sir Henry Parkes did.

20

PASTEUR'S GRAND OPENING

FOURTEEN NOVEMBER 1888. Since dawn, increasingly excited crowds of onlookers had been gathering in the streets around the new Pasteur Institute at Grenelle. A double row of policemen held back the throng, supported by troops of mounted police. Next door to the institute, spectators sat on the roof of the factory that made the l'Urbaine company's coaches. Across the road in a market garden, a ramshackle hotel that had recently been renamed the *Grand Hotel de l'Institut Pasteur* had become something of a grandstand: people hung out all its windows, others covered its roof, with some clinging precariously onto chimneys so they would get the best view of the official opening of the *Institut Pasteur* or 'the Rabies Palace' as the French press had nicknamed it.[284] Beyond the hotel, and beyond the market garden and distant wasteland bordering it, lay central Paris. Jutting above the northeastern horizon, the top of the almost completed spire of the Eiffel Tower signalled that the 1889 Paris World's Fair and its celebration of the centenary of the French Revolution were only months away.

At noon, a carriage carrying Louis Pasteur and his son Jean-Baptiste rolled along the Rue d'Utot and turned in through the institute's impressive gateway. Passing the waiting band of the

Republican Guard, whose colourfully uniformed bandsmen stood in loose formation with their instruments on the ground, the carriage pulled up at the bottom of the sweeping steps that led to the front door of the institute's main building. Pasteur, in his best suit, wearing the sash of a Commander of the *Legion d'honneur* and the decoration of Russia's Grand Cross of Saint Anne, was not yet sixty-six but he looked a frail old man as he was helped from the carriage by Jean-Baptiste. With Pasteur holding his son's arm, the pair slowly mounted the steps and went inside. From the entrance hall, past waiting colleagues and institute staff who smiled, nodded and bowed, the Pasteurs passed through the door to the left and entered the library where Pasteur would receive the official guests as they arrived.

Pasteur had named this room the Salle des Artes – Room of the Arts. This was the library that he had envisaged would feature wood-panelled walls and a mighty oak ceiling. But the struggling subscription fund could not support such extravagance so, in order to save money, the ceiling wore pink paint and the painters had used brown paint and combs to make the plaster on the walls look as if it were solid, grainy wood; their craftsmanship would fool most observers. On pedestals around this room sat busts of the better-known donors to the subscription fund, among them Czar Alexander, Count Laubespin, Madame Boucicaut, the Baron de Rothschild, and Brazil's Emperor Pedro II – who, in almost exactly one year's time (15 November 1889), would be deposed in a military coup that terminated monarchial rule in Brazil.

Twelve hundred invitations to the institute's opening had been sent out, and from 12.30 pm guests arrived in a stream that began with delegations of French science students complete with the banners of their universities. At 1.00, a fleet of carriages began depositing dignitaries at the door – first, leading members of the Institute of France; then five ministers of the government; a noted French explorer; the directors of all the leading French newspapers; and the ambassadors of Italy, Turkey and Brazil. Next, representing the Czar of Russia, came Prince Alexander of Oldenburg and the Grand Dukes Vladimir and Alexis. All were conducted to a courtyard where rows of chairs were arranged in front of a dais decorated with red velvet and gold bunting.

At 1.20, the coach of the President of France pulled up in front of the steps, with an escort of cavalry in splendid uniforms and shining helmets. President Sadi Carnot, who had replaced Jules Grévy after the Wilson scandal eleven months earlier, stepped down from the coach and came to attention. The military band struck up the *Marseillaise*, the French national anthem, and the crowd in the streets outside patriotically and enthusiastically gave voice to the lyrics. President Carnot then proceeded inside and was greeted by Pasteur. Together, Pasteur and the President adjourned to the official dais where seats were reserved for them.

Newspaper reporters were bunched to one side of the courtyard, scribbling industriously in their notebooks as they tried to depict the scene for their readers. Among those reporters was the Paris correspondent of the *Sydney Morning Herald*. 'The President, in his quiet, gentlemanly way, took the seat prepared for him, with a crowd of notables,' the *Herald* correspondent was to report. The Australian journalist went on to describe how, at a nod from President Carnot, the President of the Institute of France, Joseph Bertrand, made the first speech, 'recounting briefly the career of M. Pasteur, labouring for forty years in the cause of humanity, acquiring glory without having sought it'. Bertrand concluded by thanking Pasteur for his labours 'in the name of humanity'.[285]

Pasteur's deputy Jacques-Joseph Grancher spoke next, in his capacity as secretary of the institute's subscription fund, describing 'the struggles of Pasteur' and 'his heavy trials'.[286] Next in a line-up of speakers came Albert Christolphe, President of the Credit Foncier bank, who told amusing anecdotes about the collection of donations for the institute. The twelve hundred guests laughed, but not Louis Pasteur. Money worries still weighed him down. The institute had been constructed, it was true, and today the President of France would officially declare it open. But Pasteur was still at a loss as to how he would pay all its operating costs.

The subscription fund had by this time collected a total of 1,940,000 francs. Of this, 420,000 had gone toward purchasing the land. Another 600,000 had been set aside for the buildings.[287] But then there was the cost of equipment and, most worrying of all, the ongoing annual salaries. Pasteur had put together his 'dream team' to run the institute, including men who had loyally served him for a number of years. Emile Duclaux, now his overall right-hand man and designated successor, would be in charge of general microbiology. Joseph Grancher would run the rabies treatment service – the original reason for the institute. Emile Roux would head up technical microbiology. And Charles Chamberland had rejoined Pasteur to head production and sale of all vaccines. Pasteur had also attracted two leading scientists from Russia – Nicolas Gamaleia would run the institute's microbiological research, and Elie Metchnikov had come to Paris to take charge of the Pasteur Institute's morphological microbiology after having received a cool reception from Robert Koch in Berlin when he'd applied for a post with the German Institute of Hygiene. But how was Pasteur going to pay them all?

In coming to work at the Pasteur Institute, these eminent scientists had to forgo any university appointments or outside teaching posts; for the institute to succeed, these men must devote themselves to its work. On top of the salaries of department heads, Pasteur also had to find those of their assistants, fourteen very capable young doctors. Then there were the wages for the ancillary staff. Among those Pasteur had employed to fill the institute's more menial jobs was Jean-Baptiste Jupille, the former Jura shepherd boy who had gained fame as Pasteur's second successfully treated rabies patient; Jupille became a concierge with the institute.

To equip the institute, to meet all these salaries, and to cover the Pasteur family's living expenses, there had been approximately 900,000 francs left in the subscription fund. To add to this would be the ongoing income from vaccine sales, which were then netting Pasteur around 500,000 francs each year.[288] The balance of the subscription fund would generate an annual investment income of around 50,000 francs. With the institute's running costs in the vicinity of a million francs a year, vaccine sales and investment income combined were clearly inadequate to meet its operating costs. At this point, Pasteur's only way to meet those running costs without seeking either a bank loan or government assistance, both of which would have opened the door to loss of control of the institute, was by eating into his capital. That being the case, the subscription fund would be exhausted before two years had elapsed. The institute would then face an annual deficit in excess of half a million francs each year after existing vaccine sales had been deducted.

This looming financial crisis was the motivation for Pasteur's telegram to Loir and Germont in Australia the previous month in which he'd attempted to leverage 2,500,000 francs out of the New South Wales Government in one hit, in return for his anthrax vaccine. But his idea had backfired – Premier Parkes had not even bothered to respond to Pasteur's £100,000 'offer' and all of a sudden Minister Abigail had been too busy to discuss the vaccination of 130,000 New South Wales sheep with Loir or Germont. As a result, Pasteur's frustration with the Australians had only grown more intense. His focus had swung back to the original reason for his interest in Australia: the rabbit prize. The ship would be steadied if only the Rabbit Commission would see reason and promptly award him the 625,000 franc prize. Once he had that money he would come to some equitable arrangement regarding anthrax production in Australia.

In the courtyard at the new institute, the succession of official speeches had become so dreary that Emile Roux and his assistant slipped away before the opening ceremony had concluded, heading back to the city to resume work at their temporary Rue Vauquelin laboratory – due to budget constraints, there were as yet no facilities for them at the institute.[289] And then at last it was Louis Pasteur's turn to speak. He had prepared an address, a plea for the role of scientific research and teaching in society. But tired, emotional, and feeling the financial strain, he decided not to struggle through it, particularly now that he spoke with a slur as a result of his last stroke. Instead, he handed the address to his son, Jean-Baptiste the diplomat, and Jean-Baptiste read it to the gathering.

The *Sydney Morning Herald* reporter in the crowd was particularly impressed by a passage in that speech, where Pasteur contrasted 'the two laws which rule the world – the lust for glory that murders its victims by hundreds of thousands on the field of battle, and scientific research that is seeking to destroy the germs which shorten life'.[290] Pasteur also praised the role of critical questioning in the advancement of science. 'Always cultivate the spirit of criticism,' he said. 'Once it has been allowed to fail, there is nothing to awaken an idea, nothing to stimulate great things. Without it, nothing will hold up.'[291]

Finally, President Carnot presented the *Legion d'honneur* to Duclaux, Grancher, and also to Brébant, the generous architect of the institute, before officially declaring the Pasteur Institute open. Pasteur now took the President on a tour of the facility, and the inquisitive guests followed along behind in a flood. The reporter from the *Sydney Morning Herald*, a member of that throng, would describe the institute for his readers in detail, from the deceptive library 'richly ornamented with wood-panelling', to 'cheerful waiting rooms' in the rabies treatment section, the 'floors of gay encaustic tiles', and the garden, 'tastefully laid out, bright just now with button chrysanthemums'.[292]

And then the President, Prince and dukes departed, the government ministers, ambassadors and dignitaries took their leave, the crowds of onlookers melted away, and Louis Pasteur was left in possession of his institute, and with the worry of how he was to pay the cost of running it.

21

TRAITORS IN THE CAMP

BY THE SECOND WEEK of December, Loir and Germont had gone north to Brisbane, capital of the colony of Queensland. Everything had stalled in New South Wales since Pasteur's telegram demanding £100,000 for his anthrax vaccine. On 26 October, Loir and Germont had written Pasteur another long letter, this time explaining that the New South Wales Government no longer seemed interested in pursuing the vaccination of its stock for anthrax while Pasteur was asking for £100,000. In this letter, Loir had suggested to Pasteur that he be permitted to offer to conduct the vaccination of 130,000 sheep if their owners paid for it, the same way stockowners did in France. As soon as this letter had reached Loir's uncle in the first week of December, Pasteur shot off another telegram to Sydney: 'Your letter of 26th October received. First finish your experiments re the rabbits, as agreed upon between you and the government in May last. You are then free to vaccinate 100,000 or 200,000 sheep after agreement between you and the Anthrax Board or the Government.'[293]

Pasteur did not seem to grasp that neither the rabbit prize nor government money for the anthrax vaccine now seemed likely. But Loir did. And so he took the only offer on the table at the time.

In October, Loir and Germont had been approached by a representative of the Queensland Government – almost certainly Rabbit Commission member Dr Bancroft – with a formal proposal that they conduct research into pleuro-pneumonia in Queensland cattle. It was a subject with which Loir was very familiar. 'Pasteur had asked me to carry out experiments on pleuro-pneumonia throughout the six years that I worked with him,' Loir was to write.[294] Some years before, during a visit to a French farm where there was an outbreak of the disease, Loir had collected blood samples containing the virus, but back at the Rue d'Ulm neither he nor Pasteur had been able to cultivate the pleuro-pneumonia microbe. Now the Queensland Government was offering Loir and Germont the challenge of doing what Pasteur had failed to accomplish over the past six years.

The Queenslanders were not offering Loir and Germont a fortune – £2 ($800) a day each plus their expenses – but at least it was better than sitting around in Sydney doing nothing while they waited for the Rabbit Commission to approve the third Pasteur experiment. So the two Frenchmen had agreed. A four-man Pleuro-Pneumonia Commission was set up in Queensland on 16 November, and by early December Loir and Germont had headed off to Brisbane, leaving colonials scratching their heads as they pondered why the announced anthrax vaccination program had failed to proceed.

'For some reason that has fallen through,' the Brisbane *Courier* of 13 December reported of the program. 'On inquiry, we now learn that M Pasteur was dissatisfied with the treatment his representatives had experienced from members of the Rabbit Commission.' Pasteur, the newspaper said, 'consequently refused to allow them to proceed until a satisfactory arrangement had been come to with the New South Wales Government and himself.' The source of that information was probably Germont, who was to frequently make comment to the Brisbane press during the coming months.

Pasteur's refusal to proceed with anthrax vaccinations until the rabbit prize was settled was applauded by the Brisbane *Evening Observer*. Calling Pasteur 'the apostle of the infinitely little,' the paper observed in an editorial, 'M. Pasteur, although he is "only a scientist", evidently knows something about business.'[295]

There in Brisbane, over the Christmas period and into the new year of 1889, Loir and Germont stayed, as the welcome guests of the Queensland Government and receiving the full cooperation from officials of a newly formed Pleuro-Pneumonia Commission. The Commission included among its members Chief Inspector of Stock P. R. Gordon, whom Loir and Germont had met at the Junee anthrax trial. Over the next few months, the pair would travel inland to the Darling Downs with government officers, visiting cattle stations where pleuro-pneumonia was prevalent and taking samples from the lungs of cattle that had died from the disease.

After their first visits inland, while Germont continued to make field trips and gather samples of the virus Loir, the laboratory ace of the team, returned to Brisbane where he worked in a temporary laboratory created for him at the government immigration station in the suburb of Indooroopilly. For months on end, Loir would strive to cultivate the microbes and attenuate a vaccine. He had written to his uncle to tell him what he was now doing, and once Pasteur received Loir's letter he suddenly became excited by the idea of creating a pleuro-pneumonia vaccine. He immediately wrote back, instructing Loir on the processes he should follow in the quest.

Pasteur and his wife Marie had by this time left the Rue d'Ulm and moved into their spacious top-floor apartment at the Pasteur Institute. Having lived in rented or government lodgings all Pasteur's working life, this apartment at the Rabies Palace was the first home that he and the loyal and long-suffering Marie could claim ownership of. From his new residence, Pasteur had only to walk down a flight of stairs to several large laboratories, and with the new facilities at his disposal he now threw himself into Loir's latest Australian project.

'He set to work,' Loir later said, 'and even from afar I could detect his enthusiasm, through the instructions he was giving me.'[296] By this time, Pasteur was leaving all other research projects to his younger, fitter institute underlings and merely reviewing their work. But Loir's Australian project suddenly reinvigorated him, and for the first time in years he became obsessed, personally working in the laboratory on pleuro-pneumonia research and dictating long letters to Loir about what he was attempting, and what Loir should try. 'By every mail,' Loir wrote, 'new suggestions arrived. Despite the distance, it was an unending collaboration.'[297]

Most of Pasteur's discoveries had come after years of trial and error in the laboratory, so Loir knew not to expect an overnight breakthrough. Day after day, Loir worked on his own in the temporary Brisbane laboratory, watching the press for news of progress from the Rabbit Commission in Sydney and keeping an eye on the post for letters from France and from his many new friends in New South Wales. To calm his mother's fears that he wasn't looking after himself on the southern continent, Loir had his portrait taken by a leading Brisbane photographer, and sent it home to her. He seems not to have been lonely. His celebrity status had followed him from Sydney, and the local newspapers regularly wrote about him and Germont, generating numerous invitations to lunch and dine at the homes of the local social elite.

By this time, Loir's written and spoken English had much improved, but he was still thinking in French and then translating everything as he spoke. A local newspaper was to report that at one of the Brisbane soirées where he was a special guest he sat down at the card table to play a game of whist with his hosts. Before long, he found himself with a winning hand.

'You lucky dog!' said an attractive lady across the table from him with a smile.

A little later, that same lady was dealt a winning hand and Loir, wishing to return the compliment but thinking in French, a language dictated by gender, said with the best of intentions, 'You lucky bitch!'[298]

Throughout his career, when Louis Pasteur saw an opportunity to further his scientific interests, he grabbed it with both hands. In early 1889, still obsessed with raising money for the new Pasteur Institute, a sudden political change in New South Wales presented Pasteur with what he saw as an opportunity. As usual he set out to exploit it.

Premier Sir Henry Parkes was as astute a political player as New South Wales had ever seen, but he was taken completely by surprise when he was ambushed in the Legislative Assembly on 15 January 1889. The previous October, just after the announcement of the successful anthrax trial at Junee, Parkes had appointed the second of his three new railway commissioners: William Fehon from Victoria. Soon, a Protectionist member of the Legislative Assembly, J. H. 'Jack' Want, a forty-four-year-old lawyer and former attorney-general, was accusing Fehon of corruption in Victoria and demanding an official inquiry into his past dealings.

Sir Henry Parkes had looked into the accusations, and by December had found them groundless. 'I had promised to make inquiries,' Parkes was to later write. 'Those inquiries I made, and the government were satisfied by the result that the charges and insinuations against the commissioner were unfounded. The papers were laid before Parliament.'[299]

On 15 January, Jack Want proposed that the House should again debate the matter of Commissioner Fehon's appointment. On the other side of the House, Parkes merely folded his arms; he did not even attempt to respond for this was old cabbage cooked a second time as far as he was concerned. Civil servant Charles Lyne was to say that Parkes at this moment 'resented what seemed to him unreasonable persistency'.[300]

Jack Want, a short man who was possessed of a shiny bald head, a walrus moustache and a short temper, now became angry. 'I call upon every right-minded and honourable man in the House to support me if I call for a division on this motion!'[301]

January being the middle of summer in New South Wales, many MPs who supported the Free-trade government were absent from the parliament, which began sitting on 8 January after the Christmas break. The opposition benches seemed well populated on 15 January, unlike the government side, but as the division bells rang Parkes still felt he commanded a majority in the House at that moment. But then, as Parkes watched with surprise, a number of members of the House who had previously supported his government crossed the floor and voted with the opposition. The Parkes Government was defeated by nine votes. Next morning, as convention required, Parkes visited Lord Carrington at Government House and informed him that he no longer had the confidence of the House, and tendered his resignation as Premier. The following day, Protectionist leader Sir George Dibbs was sworn in as the new Premier.

The Protectionist ambush of Parkes seems to have been carefully orchestrated and planned weeks in advance, in collusion with a number of disillusioned Free-trade supporters. The members who crossed the floor and betrayed Parkes must have known that the vote would be brought on that day. Other Parkes supporters had also been aware that the vote was going to be called, and stayed away from the House. The reason most often given for this mass betrayal of Parkes was his increasingly dictatorial style. Even the members of his cabinet had complained that, with his massive majority in the House, Parkes no longer bothered to consult them, just pushed ahead with whatever course he wanted to pursue.

Among those who failed to support Sir Henry that day was Thomas Garrett, who had been removed from the Lands portfolio by the Premier. Garrett's defection was predictable, but the identity of many of the others who became what Parkes perceived as traitors came as a genuine surprise to Sir Henry and his closest friends. Mines Minister Frank Abigail was one of those who walked away from Parkes. Abigail had been furious that the Premier had withdrawn government involvement from the Pasteur anthrax vaccination program. Abigail had probably advocated giving Pasteur his £100,000 to permit the program to proceed, but Parkes was not going to be blackmailed by anyone. After the humiliation of losing control of his rabbit eradication competition, Abigail had seen the anthrax program as his vindication, his chance for glory, but Parkes had also denied him this. By the time the 15 January vote came around, Abigail was a bitter man.

For the next six weeks, the new Dibbs Government struggled to govern. One of its first initiatives was a royal commission into the accusations against Railway Commissioner Fehon – the royal commission would exonerate Fehon of all charges and he went on to serve as a railway commissioner for many years. By the beginning of March it was clear that the Free-traders who had allowed Parkes to be defeated were not going to support the Protectionists when money bills came up for the vote; Dibbs was forced to call a general election. Once again, Parkes went out on the political stump with the vigour of a young man. And once again crowds rolled up to his meetings.

While this election was going on, Louis Pasteur in Paris received a letter from Loir in which he told of the change of government in New South Wales. Pasteur had come to blame all his problems with the Rabbit Commission on the Parkes Government, which had created it. To Pasteur's mind, the new government of Sir George Dibbs must surely be easier to deal with if it opposed Parkes. It seems to have occurred to Pasteur that he might be able to cajole this new government into allowing him to carry out the third chicken-cholera test in open country, and so prove that he should be awarded the rabbit prize. To do this, he would have to be more cooperative on the subject of the anthrax vaccine program.

So, in the second week of March, Pasteur dispatched several suitcases to Sydney containing tubes of anthrax vaccine – exactly how many tubes is unclear, but they would have numbered in the hundreds, with enough vaccine in each tube to vaccinate one to two hundred sheep. On 12 March, Pasteur sent a telegram to Loir and Germont, telling them that he had consigned the last anthrax vaccine suitcase to them the previous day, but that the vaccine was only 'to be used after third successful test on rabbits'.[302]

This telegram was retransmitted by the Colonial Secretary's office to Germont and Loir in Brisbane. When Loir read this latest message from Pasteur he would have thrown his hands in the air. Unlike his uncle, Loir knew that the Sydney government, no matter what its political complexion, had absolutely no power over the Rabbit Commission. Yet Pasteur was sending him vaccine that had a potential market value of 3 francs for each two-dose vaccination – the entire shipment was potentially worth hundreds of thousands of francs. It was a crazy gambit, one that Loir knew had no chance of succeeding. The Rabbit Commission would do what it wanted, irrespective of what was done with Pasteur's anthrax vaccine.

In New South Wales, the latest election campaign created great interest. Yet the public enthusiasm for Parkes and his agenda that had manifested itself at the last election wasn't there. Some voters accused Parkes of not implementing the free-trade policies they had expected of him. Others accused him of being dictatorial in government. There was also a personal factor that weighed on the minds of some more stuffy voters: in February 1888, Parkes' first wife Clarinda, from whom he had been estranged for some time, passed away. Twelve months later, on 6 February 1889, when the election campaign was unofficially already under way, seventy-three-year-old Parkes married his mistress Nellie Dixon, repaying her loyalty and devotion, and in the process making her Lady Eleanor Parkes.

'The little affair was kept very quiet,' remarked the Sydney *Bulletin* of the wedding. In front of just a handful of witnesses, it was conducted at St Paul's Redfern by the Reverend B.W. Boyce, a close friend of Sir Henry, with Sydney Burdekin, a Free-trade parliamentarian, as best man. 'Next day Parkes was on the stump in the Dubbo district,' the *Bulletin* reported. 'What a wonderful old man, and *what* a way to spend a honeymoon!'[303]

This marriage was condemned by some colonials because Parkes had married a woman young enough to be his granddaughter and because she was a divorcee, but mostly because he had only waited twelve months since the death of his first wife before remarrying – an unseemly rush to the minds of some critics. Two of Parkes' daughters left home and refused to have anything to do with Nellie, while the third bemoaned the fact that her stepmother was younger than she was. The new Lady Parkes 'is of an unobtrusive and retiring disposition', commented the Sydney *Truth*, 'and unless engaged in some charitable movement, is rarely seen at any public function'.[304] Nellie was never received at Government House, and Sir Henry's future visits there would always be solo affairs.

Yet overall, Parkes' popularity as a political figure only dipped a little, for a majority of people in New South Wales still saw him as by far the best leader for the colony. When the votes were counted following the March election, Parkes and his loyal core of parliamentary supporters had been reelected. But those who'd betrayed him, as well as those who had, in the words of Charles Lyne, been elected in 1887 by simply attaching themselves to Parkes' coat-tails and professing to support him only to later desert him, were punished for their infidelity, and punished severely.[305]

The Free-traders were returned to power, but only after Parkes cobbled together a coalition which gave him a majority of just three seats. On 17 March Sir Henry Parkes was once again sworn in by Lord Carrington as Premier of New South Wales and formed his fifth ministry. Not one of the new ministers that he appointed to his 1889 cabinet had served in his last ministry. As usual, all cabinet ministers appointed by Parkes were members of parliament even though the New South Wales Constitution, like the American system, allowed the Premier to select cabinet members from outside parliament. Charles Lyne considered this latest ministry one of the colony's best ever.[306]

Come the election, Frank Abigail had joined forces with his former enemy Thomas Garrett and four other former Parkes supporters, running under the banner of a new political entity, the Liberal Party. All six were soundly defeated. Frank Abigail and Thomas Garrett both lost their seats. Garrett would be dead within months, and the party he founded sank without trace. Another supposed Parkes supporter who was trounced in this election and lost his seat was Dr Camac Wilkinson, member of the Rabbit Commission and chairman of its Experiment Committee. Perhaps the voters of Glebe were aware that Wilkinson had been working behind the scenes against Sir Henry Parkes, and he paid the price for it.

Wilkinson was stunned by his defeat. He would not have even contemplated losing. Now, even though he had committed to duties at Sydney University in 1889, including serving on a committee that was to build a pathology museum, the young doctor threw up everything and bought a passage to England on an Orient Line mail steamer. He would sail for England as soon as the Rabbit Commission's Progress Report was handed in.

Not that Wilkinson had neglected his role in the Rabbit Commission's deliberations. The Commission had met for the last time on 16 October the previous year, after which all commissioners had returned home. By that time Queensland's Dr Joseph Bancroft had ceased to attend sittings, and his seat was occupied for the last two meetings by Henry Tryon, the thirty-one-year-old Queensland Government Entomologist and Assistant Curator with the Queensland Museum. Privately, the commissioners had agreed that a Rabbit Commission Progress Report would be issued in April the following year, a timing that would coincide with the anniversary of the arrival in Australia of the Pasteur Mission, with a final report to be issued later in 1889.

Wilkinson had already written his Experiment Committee's Report for inclusion in the Progress Report, and had worked with Professor Harry Allen on the wording for the Progress Report's preface. By the time Sir Henry Parkes was sworn in as Premier once more, on 17 March, and by the time Louis Pasteur's last anthrax vaccine suitcase was on its way to Australia, the Rabbit Commission had already committed its verdict on Pasteur's rabbit eradication competition entry to paper.

At the same time, an item was published in the Brisbane press headlined, 'Traitors in the Camp'. 'It would appear that there are "traitors in the camp" of the Rabbit Commission of New South Wales', said the news story. According to the paper, a letter had appeared in an eminent English medical journal, the *Medical Record*, written by an unidentified Sydney doctor connected with the Rabbit Commission – the implication being that this was Camac Wilkinson as he was the only Sydney doctor still connected with the Commission. And in that letter, the doctor had revealed the Commission's decision on Louis Pasteur's competition entry well before it was to be officially presented to the government in Sydney. According to the anonymous medico, writing in the *Medical Record*, the decision was a negative one.[307]

Lending credence to the suggestion that there were leakers in the Commission camp, just before Christmas a brief item had appeared in the Sydney press reporting that a source close to the Commission in Melbourne had revealed that the Rabbit Commission's Progress Report would recommend against the Pasteur method.[308] This was months before Dr Katz's own experiments with chicken-cholera would be completed – his last Rodd Island experiment took place in March 1889. Back then in December, there had been no reason to believe that the report out of Melbourne had come from a credible source. But now, in March, this latest 'inside' report from England pointed to the Rabbit Commission leaking like a worm-ridden boat. It also suggested that the judges had made their decision even before the scientific jury had reached its verdict.

22

THE PROGRESS REPORT

IN THE LAST HALF of March, Adrien Loir, working on pleuro-pneumonia research in Brisbane, learned that the Rabbit Commission was expected to hand down its promised Progress Report in early April. He and Germont, who stayed at Brisbane's Queensland Club from 12 March and then the Union Club from 22 March, prepared to return to Sydney so they could be on hand when the report was released.

On 20 March the pair submitted an interim report on their pleuro-pneumonia work to Queensland's Pleuro-Pneumonia Commission. In that report, Loir promised to continue the research on pleuro-pneumonia at the new Pasteur Institute in Paris, in conjunction with Pasteur, from where a fuller, more scientific report would be issued.[309] To make such an undertaking, and to commit himself to a return to Paris in the near future, Loir had clearly made up his mind by then that the Rabbit Commission's upcoming Progress Report was going to deliver a negative verdict on Pasteur's rabbit competition entry. The two discouraging reports in the press from apparent Commission insiders seem to have influenced that attitude.

Also in his pleuro-pneumonia report, Loir recommended that the Queensland Government set up a stock institute in Brisbane to conduct research into animal diseases. At the suggestion of its Pleuro-Pneumonia Commission, which was delighted with the work of Loir and Germont, the Queensland Government quickly presented Loir with a cheque for £1000 ($400,000), in thanks for the work that he and Germont had done in Queensland and for the work they had undertaken to continue in Paris. This was on top of the £2 a day they had already been paid, which would have totalled some £90 ($36,000) each. Loir gratefully accepted the big cheque, worth 250,000 francs, which he would donate to the Pasteur Institute subscription fund.

By 2 April, Loir and Germont were back in Sydney, where a reporter from the *Daily Telegraph* tracked them down. Loir reminded the newsman that the Pasteur Mission had asked the Rabbit Commission for permission to carry out Pasteur's open country experiment the previous 4 August and were still waiting for an answer. When the reporter asked why the Pasteur team had not proceeded with the planned anthrax vaccinations, Loir replied that they could not proceed with the anthrax matter until the rabbit question was settled.

'Is it that you want the £25,000 bonus?' asked the reporter.

'No, that is not it,' Loir responded. 'All we want is an opportunity to complete our experiments.'

'What do you propose to do now?'

'We must ask the government for fair play.'

'And if you do not get it?'

'We shall return to France.'[310]

By the end of March, with Dr Katz's Rodd Island experiments on behalf of the Commission having been wrapped up on 9 March, the Rabbit Commission's Progress Report had been printed and was ready for presentation. On 2 April, Professor Harry Allen travelled to Sydney from Melbourne and new Queensland rabbit commissioner Henry Tryon came down from Brisbane. On the 3rd, Allen presided over the Commission's final meeting, which was also attended by Tryon, Camac Wilkinson and Alfred Dillon Bell. All four applied their signatures and seals to the Progress Report and Secretary Hugh Mahon signed on behalf of the absent commissioners. That same day, Professor Allen formally handed the report to Sir Henry Parkes' new Minister for Lands, James Brunker, and also released it to the press. The government forwarded a copy to Loir and Germont.

This Progress Report, which ran to many pages, included minutes of Commission proceedings taken at sittings in three colonies and involving thirty witnesses, plus all of Dr Katz's detailed experiment reports. It would be the preface on which most readers would focus, with its summary of all the schemes proposed in response to the advertised competition, and the Commission's findings. The report dismissed the majority of competition entries in a few paragraphs (see Appendix for more details of the other schemes and of Dr Katz's experiments). Most space was reserved for Pasteur's method.

Early in the preface there was a section headed, 'Difficulties With M. Pasteur's Representatives'. This section highlighted Loir and Germont's refusal to undertake the experiments proposed by the Commission after Frank Hinds had seemingly agreed that they would do so. In a later section, 'Further Negotiations With M. Pasteur's Representatives', it was noted that Pasteur had consented to an additional experiment to demonstrate the contagious nature of chicken-cholera among rabbits, but had declined to provide chicken-cholera microbes for the Commission's own experiments until his own had been concluded. The Commission's ultimatum to the Pasteur Mission was then outlined. At no point did the Progress Report explain how, after the issuing of this ultimatum, Pasteur's team suddenly came to conduct their own experiments on Rodd Island anyway, in defiance of the ultimatum. The report merely stated, untruthfully, that negotiations had 'advanced a stage', intimating that this advancement had permitted the Pasteur experiments to go ahead.[311] No reference was made to the intervention of Sir Henry Parkes, an incident that had been a blow to the commissioners' pride.

A little further on in the report, Pasteur's letter to Chief Inspector of Stock Alexander Bruce was reprinted, followed by scathing criticism by the Commission. 'M Pasteur, however, seems not to understand the terms of the proclamation concerning the reward,' the report observed, before mentioning Pasteur's reference to a 'secret' relating to his method, a secret that he had said was known only to him. 'M. Pasteur's letter [to Bruce] implies that he expects the reward to be adjudged to him for his scheme on such proofs of its efficacy as he may determine, and before the mode of carrying it out on a large scale shall be revealed.'[312]

Of the experiments carried out by Germont and Loir, the Progress Report said: 'The experiments of demonstration, carried out to the instructions of M. Pasteur, were regarded by the Commission as unsatisfactory. But it was now possible to institute a new series of tests which would fully try the real issue.'[313] Oscar Katz's experiments were then outlined. His recommendation that further experiments be carried out to determine whether chicken-cholera already existed in Australasia was included in the preface. It was the only recommendation from Katz to be given space there.

On the positive side, the Progress Report did note that as a result of both the Pasteur and Katz experiments it could state that domestic animals were not harmed by chicken-cholera and that the possibility of the widespread dissemination of the disease from rabbits to birds was an 'exaggerated notion'.[314]

But it was the effect of chicken-cholera on rabbits that most concerned the report's authors. The report concluded that the experiments had shown that while rabbits were easily killed by the addition of chicken-cholera microbes to their food, 'the disease does not spread freely from infected to healthy rabbits'.[315] This conclusion was incorrect, as anyone who bothered to read Katz's detailed data at the back of the report would have discerned. In both the Germont–Loir experiments and the Katz experiments on Rodd Island, the disease had been shown to be lethally contagious among rabbits – via contact in hutches with the droppings of rabbits that had contracted the disease, as well as in hutches in the open, and in burrows from contact with rabbits that had died from the disease.

Just how contagious had proved to be the contentious point, a question whose answer was muddied by the death of large numbers of test rabbits from starvation. The report preface failed to make clear that these deaths came about as a result of the Experiment

Committee persisting in bringing wild rabbits onto Rodd Island from the Hay district against the advice of commissioner Dr Bancroft. In the Commission's minutes, Bancroft is seen to warn that wild rabbits would not take feed in captivity and would thus starve to death. For this reason and others, Oscar Katz had recommended that further experiments be carried out in the open, as Pasteur required, to fully test the contagious nature of the disease in natural conditions. But the Progress Report's preface made no mention of Katz's reservations nor of his recommendation in this respect.

The Progress Report even misrepresented Katz's findings. It said: 'Dr Katz indicated that while the microbes retain their virulence for a time when mixed with putrefying matter, there is a limit to their power of survival.'[316] What Katz had actually said was that the microbes of chicken-cholera retain their virulence in the body of a dead rabbit even as it putrefies, and remain virulent for 'a considerable time' after the death of a rabbit, in the putrefied matter.[317] Katz had made no reference to 'mixing' microbes with putrefying matter, a term that implied the physical act of spreading microbes by hand. At the same time, the report's comment that 'there is a limit to their power of survival' is an obvious truism – nothing lasts forever – but it was skilfully introduced into the report to imply a very limited life to chicken-cholera microbes in putrefying matter. Katz had in fact reported just the opposite.

This careful manipulation of words was evident elsewhere in the Progress Report. Several times the report made reference to the fact that test rabbits 'died during the experiment from causes independent of chicken-cholera'.[318] While a reading of the full report would show that these deaths had resulted from the Experiment Committee's use of rabbits that were already unhealthy, or which died from shock while in captivity, to the uninitiated reader this would imply that chicken-cholera was a failure.

Adrien Loir seems to have been expecting the worst, so it must have been no surprise to him that the Progress Report summed up: 'Generally, therefore, it appears that the destruction of rabbits on a large scale can only be obtained by feeding the rabbits with the microbes of the disease.' This inaccurate observation was followed by the claim that chicken-cholera was therefore no better a remedy than the poisons currently in use.[319]

The report concluded: 'The Commission cannot recommend that permission be given to disseminate broadcast throughout Australasia a disease which has not been shown to exist in these Colonies, which in other countries prevails in disastrous epidemics, among fowls, but which has never been known to prevail naturally among rabbits.'[320] Here the Commission was stating that chicken-cholera had not been shown to already exist in the colonies, and stressed the potential for 'disastrous epidemics' if it were allowed to be introduced. This again ignored a recommendation from Dr Katz and misconstrued his report. He had only been able to test nine chickens said to have died from chicken-cholera in New South Wales. He had reported that while he'd been unable to detect chicken-cholera in those birds, it was essential that more extensive experiments be conducted over a long period of time before any firm conclusion could be reached about whether or not chicken-cholera already existed in the colony.

The Progress Report's final word on Pasteur's chicken-cholera rabbit eradication method was this: 'The Commission, in fact, does not feel justified in recommending any further expenditure by government in testing the efficacy of this disease.'[321]

All eleven commissioners signed off on this conclusion, and on a recommendation that the government legislate to force landowners to instal rabbit-proof fencing to counter the rabbit plague. But there was an addendum. Six of the eleven commissioners agreed that, merely as a courtesy to Monsieur Pasteur, if he wished to carry out his large-scale chicken-cholera experiment on some sheep station, he should be allowed to do so – but at his own expense, with no involvement from the Commission or the government, and with the results of any such experiment being ignored by the Commission.

Commission President Professor Harry Allen, who was one of five commissioners who voted against allowing Pasteur to conduct this trial, was to comment: 'The Commission is of the opinion that no advantage can be gained from such an experiment, other than that M. Pasteur's representatives would not be able to complain that any of the demonstration which they were sent to perform had been prohibited by the government.'[322]

The Commission, in finding that none of the entries in the rabbit eradication competition was acceptable, made it obvious that it had no intention of permitting the introduction of a biological rabbit remedy of any kind, and no intention of allowing the New South Wales Government award the £25,000 prize to anyone.

The press of the colonies reported the Commission's findings word for word, and took them at face value. 'In one sense the result of this extensive and laborious investigation is a disappointing one,' said the Sydney *Daily Telegraph* of 4 April. But the disappointment expressed by the *Telegraph* and other papers was merely a regret that the competition had not generated the solution to the rabbit plague that so many had been hoping for. Not one paper expressed any disappointment with the Commission itself, or looked deeply or questioningly into its Progress Report.

Not a solitary editor or reporter looked beyond the Progress Report's conclusions. Neither did MPs when the Parkes Government tabled the report in the New South Wales parliament. No one noticed that both Harry Allen and Camac Wilkinson, the authors of the final report, had signalled the Commission's outcome on the very first day it sat, back in April 1888; yet it was all there in the minutes. No one spotted that the Progress Report either misread or misrepresented Dr Oscar Katz's reports, or that five of the commissioners had gone against the recommendation of their own Chief Expert Officer regarding a large-scale inland trial of chicken-cholera.

No one questioned the Progress Report's key recommendation – that the government legislate to make rabbit-proof fencing obligatory for all rural landowners. Yet they only had to look at the minutes of the Commission sitting of the previous 26 April, just ten days after the Commission met for the first time, to see wire mesh-manufacturing commissioner Edward Lascelles tell his colleagues: 'The Under Secretary for Lands has stated in a letter just received that the regulation netting adopted by the government will cost £38 per mile. Now, I am prepared to sell wire which, I contend, will be equally as effective, at the rate of £21 per mile.'[323] At no time had Lascelles felt compelled to stand down from the Commission because of his obvious conflict of interest, nor did any of his fellow commissioners suggest he do so.

Unlike Lascelles, another commissioner did experience a sudden attack of conscience over a conflict of interest. Fellow Victorian Albert Pearson, another strong promoter of rabbit-proof fencing, announced at the 16 October sitting of the Commission, its last full sitting, that he had invented 'certain improvements in wire-netting fences'.[324] Pearson had then offered to resign should his colleagues feel that this might compromise his continued membership of the Commission. Fred Dillon Bell had immediately moved that the Commission's interests would be more seriously harmed by Pearson leaving than if he remained, and that his resignation should not be accepted. The motion was carried, and Pearson had continued as a member of the Rabbit Commission. This was in the Commission's minutes for anyone to read, but it seems that no one did.

There was yet another reason why Pearson should have been disqualified from sitting, a reason he only made public on the day the Progress Report was delivered to the government. Pearson revealed that he had an interest in poultry farming – not surprisingly, poultry farmers were totally against the introduction of chicken-cholera into the colonies. Adrien Loir was to learn that Pearson was in fact the president of a poultry-breeders association.[325]

Yet no eagle-eyed reporter or parliamentarian noticed or objected to the fact that Lascelles and Pearson stood to gain from the recommendations of the Commission, nor that the Commission had been prepared to give another of its members, Quin, 10,000 acres of free rabbit-proof fencing. No one was aware that the Commission had concealed the fact that one of its commissioners, Alfred Dillon Bell, had also been on the Commission's payroll as Dr Oscar Katz's experiment assistant, which put Bell in a position to influence the outcome of experiments rather than simply be an independent observer of them.

No one picked up that both Allen and Wilkinson were prejudiced against all biological methods due to the influence of German anti-vaccinists led by Pasteur's arch-rival in Germany, Robert Koch.

Nor was anyone aware that Camac Wilkinson and the Experiment Committee had secretly sought and obtained chicken-cholera and rabbit septicaemia microbes from Robert Koch in an attempt to thwart Pasteur. This was yet another fact hidden from the government, from parliament, from the public, and from Louis Pasteur by the authors of the Rabbit Commission's Progress Report. And while the minutes at the rear of the Progress Report did reveal that those other microbes had suddenly arrived from Europe, no one asked how, or who'd supplied them, or why.

On the other side of the ledger, every newspaper and every politician accepted the Commission's unsubstantiated statement that chicken-cholera did not already exist in the colonies, and believed the report's claim that to use it would run the risk of 'disastrous epidemics'. Fear of the unknown drove a stake into the heart of Pasteur's remedy.

It has to be said that, all in all, Dr Oscar Katz had been a relatively impartial judge of the chicken-cholera experiments. This was despite his German training, which made him the natural adversary of the Pasteur method, and despite an abrasive manner and limitations as a microbiologist. Most importantly, Katz had recommended that trials with rabbits and chicken-cholera be extended to the next stage, in open country, as Pasteur wished. And he had recommended that much more experimentation take place before any conclusion be drawn about whether or not chicken-cholera already existed in Australia.

On the other hand, the Rabbit Commission itself had not been impartial and had acted improperly and deceitfully. Its findings were faulty and deliberately skewed against Pasteur. Several of its members had serious conflicts of interest and, from the outset, most members were driven by personal agendas. The two men chiefly responsible for the writing of the Progress Report, President of the Commission Professor Harry Allen and Chairman of the Experiment Committee, Dr Camac Wilkinson, both supporters of the German anti-vaccinist position, clearly had no intention of allowing Louis Pasteur to gain a microbiological foothold in Australia or New Zealand.

For all this, with the Rabbit Commission's final report scheduled to be handed down in October, the last shots in the rabbit war had yet to be fired.

23

AN INTERNATIONAL INCIDENT

WHEN LOIR AND GERMONT arrived back in Sydney from Brisbane at the end of March, just prior to the release of the Rabbit Commission's Progress Report, they found themselves in the middle of a political storm. Before the pair went to Queensland, they had told their friend Arthur Devlin about how their letters had been intercepted, opened, and apparently read by the Colonial Secretary's office. Devlin, who seems not to have been a supporter of Sir Henry Parkes, decided to weigh into the election campaign then raging, and on 8 March wrote a letter to the editor of the *Sydney Morning Herald* in which he complained that Pasteur's representative had not received 'fair play' in New South Wales, and in which he put a case for Loir and Germont to be permitted to conduct the final Pasteur experiment in open country. In support of his claim of foul play, Devlin cited the intercepted letters and the 'mislaid' telegram of 18 July.[326]

During the remaining days of the election campaign Sir Henry Parkes did not say a word about Devlin's letter or the accusations it contained. So a week later, with Parkes returned to government for the fifth time, Legislative Council member Dr John Creed, the man who'd originally suggested that a prize be offered for a biological cure to the rabbit plague, wrote to the *Daily Telegraph* in support of

Devlin's letter. In addition, Creed claimed that a letter sent to him the previous August from Agent-General Cooper in London had similarly been intercepted and withheld by Premier Parkes.

Creed's letter to the press prompted an immediate response by Principal Under Secretary Critchett Walker, who told the papers that the opening of letters to Loir and Germont had been an accident, while the delay in delivering Louis Pasteur's 18 July telegram was 'purely an oversight', and that 'these gentlemen appeared to have been satisfied at the time with the explanation given'.[327] Creed countered at once, decrying Walker's response as inadequate and commenting, 'The French are particularly sensitive as regards anything which concerns the illustrious M Pasteur, of whom they are justly proud, and it is not out of the range of possibility that the matter might become an international question.'[328] And so it did.

Through the last week of March and into April, Creed and Devlin bombarded the press with letters, expressing dissatisfaction with Critchett Walker's excuses, forcing Walker to issue a statement on 29 March. In this statement, Walker summed up that, as Loir and Germont had expressed satisfaction with the answers they had received from him the previous August, 'it is clearly their business alone to make a further representation in the matter'.[329] Loir and Germont, having only just returned from Brisbane, read Walker's comment about this matter only being their business, and feisty François Germont took it as an invitation to enter the fray.

Germont promptly took Loir to see Dr Creed so that they could present the Pasteur Mission's side of the story. 'Messieurs Germont and Loir have now returned to Sydney,' Creed informed the *Daily Telegraph* in a letter published on 1 April, 'and I am informed are prepared to go thoroughly into the whole of the circumstances when the necessary opportunity is afforded them.'

That same day, Germont wrote a long letter to the *Sydney Morning Herald* in his name and Loir's name, detailing the interception of the letters and telegram. He took particular exception to a claim by Critchett Walker in his 29 March statement that his office had been unable to locate Germont and Loir in the first half of July to enable delivery of the box of anthrax vaccine. 'This is a very strange assertion,' Germont wrote. 'Mr Critchett Walker could not experience such a great trouble to find us as he mentions, as he was living in the same house in the same flat with one of us.' Germont then went as close as he dared to exposing the relationship between Walker and Hinds by mentioning that there was a connecting door between their bedrooms, then stating that, 'we have special reasons to believe that Mr Walker did not ignore our presence in that house'.[330]

Germont's implication that an improper relationship had existed between Walker and Hinds – improper according to the social standards of the time – was not picked up by either the press or the opposition. It was soon established that Walker had been at the Centennial Exhibition in Melbourne at the time of the box delivery episode, taking him out of the firing line. Besides, not only was Walker well-respected in parliamentary circles, he was in a position to have known about the peccadilloes of members of parliament, so it could have been dangerous for MPs to have antagonised him.

In the 1 April letter, Germont also pointed out that the delay of the 18 July telegram from Pasteur had meant that he and Loir had proceeded with the anthrax experiments in ignorance of Pasteur's instruction that they refrain from those experiments until the rabbit prize was settled. Germont concluded by revealing that he and Loir had not been satisfied with the explanation received from Critchett Walker in Sir Henry Parkes' office the previous August. This letter from the French scientists was to guarantee that the matter of the interception of their correspondence would be taken further.

The colonial press covered the affair with a mixture of horror and delight. The Brisbane *Courier* headlined the story, 'A Serious Charge',[331] while the *Clarence and Richmond Examiner* led with 'A System of Espionage'.[332] Meanwhile another paper wrote, tongue partly in cheek, 'To put it mildly, Messieurs Germont and Loir have got the Principal Under Secretary into a corner . . . Imagine the courtly and official Mr Critchett Walker facing two elaborately polite but exasperated *Français. O, ciel.*'[333]

On 3 April, William Traill, a Protectionist member of the Legislative Assembly, laid sixteen questions before the House relating to the affair. Sir Henry Parkes advised that he would later make a detailed written response to all these questions, but remarked that he had never seen any of the letters addressed to Germont and Loir. As for the 18 July telegram, he told the House: 'The "HP", I imagine, must have been written by me, and it is exactly the thing I should write. If the telegram had been mislaid, I should write "mislaid" upon it, no doubt.'[334]

When a week passed without the Premier providing any further answers, the opposition launched simultaneous attacks in both Houses. On Thursday, 11 April, as soon as the Legislative Council sat at 4.30 pm, Dr John Creed stood and moved that an official inquiry be held into the alleged detention by Premier Parkes' office of correspondence addressed to Germont and Loir. Creed informed the House that it was 'a matter of indifference' to Germont and Loir whether such an inquiry took place in New South Wales or in Britain. He quoted a letter from Pasteur to Germont and Loir in which Pasteur had said, 'Lord Salisbury has been fully informed by Lord Lytton concerning what has happened to us in Australia.' In this letter, Pasteur had gone on to say that he was awaiting the results of his complaints to the British Government. It would later transpire that Pasteur had also complained to the Prince of Wales, with whom he was on friendly terms.

'I suggest,' Dr Creed said to the House, 'that an inquiry be held by "some high judicial authority."'[335]

George Simpson, a Free-trade Legislative Council member and Parkes' new Attorney-General, gamely defended his Premier against Creed's barrage of questions and insinuations. The battle between Creed and Simpson swung back and forth across the House for several hours before Creed's motion for an official inquiry into the interception of Pasteur's correspondence was finally put, and defeated.

Meanwhile, an attack in the Legislative Assembly had been launched at precisely the same time as that in the Upper House. Protectionist member William Traill, who had directed the sixteen questions to Parkes eight days earlier, launched his next salvo. The black-bearded, cigar-smoking former editor of the Sydney *Mail* and the *Bulletin*, Traill focused on the slur that the opposition claimed this affair must surely bring to the international reputation of New South Wales.

After a lengthy harangue from Traill, Sir Henry Parkes slowly came to his feet. Members and gallery waited expectantly for his first comments on the affair as, with right hand hooked onto the lapel of his jacket, the septuagenarian Premier surveyed the House. 'I do not think that this country is in much danger of having the finger of scorn pointed at it arising out of this transaction,' he said with a wry smile, bringing chuckles from his supporters. 'Nor do I think that there is any danger of any slur being applied to it. It may be a very appalling thing indeed if this French gentleman has communicated his views to the Prince of Wales. That may possibly overwhelm us. But I do not think that it will injure us very much.'[336]

Parkes then proceeded to deal with each of the items of correspondence involved in the allegations, one at a time. He had genuine excuses for all the items bar one – the 18 July telegram that he had personally withheld. Critchett Walker, who had consistently maintained that the Premier had never seen any of the Pasteur correspondence, had always quickly brushed over the matter of the telegram, but Parkes had yet to explain why he had written 'Mislaid H.P.' on it. Now, in the House, the Premier said, 'I know hardly anything about that telegram, and I could not say at this moment whether or not I wrote that word "Mislaid," and the letters "HP".'[337]

In defending himself, Sir Henry could have pointed out that, far from acting against Pasteur, he had helped him and his representatives by covering the high cost of their telegrams; but he didn't. He could have revealed that he had also worked on behalf of Pasteur and his representatives behind the scenes, ensuring that Pasteur's experiments were carried out by his own representatives on Rodd Island. But he also chose to keep this secret. He could have destroyed sympathy for Pasteur by announcing that the French scientist had attempted to blackmail the colony out of £100,000 in return for his anthrax vaccine, but Parkes didn't pursue this course either. He knew that he had a majority in the House, if a slender one now, and all he had to do was bluff his way through.

Despite the discrepancy in Parkes' story regarding the telegram, a telegram about which he claimed to know nothing, all the other incidents were satisfactorily explained away as the bureaucratic blunders they had genuinely been. Parkes, a crafty parliamentary performer of thirty-five years' experience, reigned supreme, like a wise old headmaster in a school full of ruffians and bright but naïve boys. Traills' motion for an inquiry into the Pasteur correspondence affair was defeated along party lines.

Four days later, John Creed doggedly resumed the attack in the Legislative Council, laying on the table the intercepted letters in question which, he said, had been given to him at his request by Adrien Loir. Creed again pursued the matter of an official inquiry, and again Attorney-General Simpson defended the Premier, supported by several other members including a former Premier, Sir William Manning. Again, the motion for an inquiry was defeated. And there the matter ended. Neither Pasteur, the French Government, the British Government, the Prince of Wales nor the British Governor in New South Wales pursued the affair any further.

As for Loir and Germont, with their lease at Albany Chambers running out in May, they packed their bags and booked their passage home. By the second half of May, they had departed Australia's shores, after thirteen very eventful months and a preliminary outcome to the Rabbit Commission's deliberations which Louis Pasteur was describing as 'odious' and 'abominable'.[338] Before long, Loir would be back in Paris and living at the Pasteur Institute with his uncle and aunt and giving his address as 25 Rue d'Utot. But his departure from New South Wales would only be temporary. Within a matter of months, Loir would be back. For, despite the cheque for 250,000 francs that Loir took with him back to Paris, Pasteur's money worries had yet to be fully allayed. Nor had the rabbit question been resolved to his satisfaction.

24

BACK TO SYDNEY

IN THE LAST DAYS of the northern winter of 1889–90, Emile Légrand, a reporter with Paris's best-selling daily newspaper *Le Siècle*, made his way to the suburb of Grenelle, to the Rue d'Utot and the new Pasteur Institute. His assignment was to interview the illustrious Louis Pasteur, and to find out why Pasteur had not been awarded the rich Australian rabbit competition prize.

The previous October, the Rabbit Commission had lodged its final report in Sydney, copies of which had only reached Paris by steamer in December. In that final report, the Commission had said that it saw no reason to alter the recommendations of its April Progress Report. Yet even now, months later, the New South Wales Government had not made any announcement regarding the prize. To the French, this meant that the competition had not been concluded, and signalled that there was still hope of Pasteur being awarded the prize.

Pasteur was more than willing to talk to the reporter about the competition. The failure to secure the prize the previous year had both annoyed Pasteur and befuddled him. Why then, the reporter asked, when Pasteur's representatives had done everything that he had asked of them, had they met such stubborn resistance in Australia?

'They clashed with the malevolent intentions of the Commission appointed by the Australian government,' Pasteur told Légrand, 'for reasons which escape us – maybe scientific, more probably through pure egotism, and so as not to award the prize to a foreigner, producing all sorts of hindrances to the experiments which Monsieur Loir, my nephew, and one of my other assistants, wanted to attempt.' Pasteur could only sigh and shrug a Gallic shrug. 'Finally, after assuring themselves that all hope was gone, my *préparateurs* withdrew and returned to France.'[339]

Pasteur took Légrand to one of his new laboratories where he graphically explained the use of chicken-cholera as a rabbit destructor. 'You can see,' said Pasteur to the reporter, 'my method is not experimental. Consequently, it is impossible not to acknowledge the practicality of its worth which, scientifically, is incontestable. If prejudice can be ruled out, then the Australian government cannot fail to give the announced prize.' He shrugged once more. 'But I have never again dreamed of securing it. As for the rest, it is for the Commission to judge as it sees fit. My concern now is for the inoculation of humans, children in particular.'[340]

Later when Pasteur, walking slowly and with the aid of a stick, escorted Légrand through the busy institute on the way to the front entrance and the journalist's waiting taxi-cab, Légrand asked Pasteur if he thought the New South Wales Government would continue with the competition, and whether the medical authorities in Australia would allow Pasteur to conduct the open country chicken-cholera trial that he'd been proposing for so long.

Pasteur shook his head. 'One doubts the continuance. They fear to look to Paris and be denounced by some of their compatriots in English medicine.' No, he said, as they passed through the Pasteur Institute's grand entrance hall to the front door, he had given up on the dream of winning the rabbit prize. 'I dream, on leaving this institute that hides in an isolated corner of Paris, of the invincible influence of the French genius – a beneficial and humanitarian genius, whose glory will never be eclipsed.'[341]

If Légrand's interview with Pasteur, which was published on 12 March under the headline of, 'The Rabbit Destruction Story –

Malevolent Intentions of the Australian Commission,' had given the newsman the impression that Pasteur had lost all interest in Australia, he would have been wrong: six weeks later, Adrien Loir was on his way back to Sydney to renew the Pasteur campaign.

Sir Henry Parkes had to tread carefully as far as Louis Pasteur and Adrien Loir were concerned. With a slim majority of just three in the Legislative Assembly since the last election – and the support of some members questionable – Parkes could no longer be either dictatorial or adventurous. And if he were to overrule the recommendations of the Rabbit Commission, he could be accused of being both. So, to comply with the recommendations of the Commission, a Rabbit Bill was introduced into the new parliament to 'facilitate and encourage the erection of rabbit-proof fencing' among other things.[342]

Yet when it was proposed to the Premier that Adrien Loir be offered the use of Rodd Island for the production of Pasteur anthrax vaccine and for disease research, at no cost to Loir or the Pasteur Institute, the pragmatic Parkes readily agreed. The Rodd Island facility was sitting empty and unused, after all. This proposal seems to have originated with Loir's friend Tiger Bruce, through the latest Secretary for Mines, Sydney Smith. Premier Parkes, aware now that the Queensland Government wanted to secure Loir's services, would have guessed that the offer of a readymade laboratory and quarters free of charge, should be enough to lure Loir back to Sydney. And such proved the case.

Loir landed in New South Wales in June, this time alone. His compatriot Germont, while a competent microbiologist, had been much too fractious for this Antipodean posting. Success in Australia would depend on the sort of charm, finesse and political sensitivity that Loir possessed in abundance. So, while Loir resumed work for Pasteur in Australia, Germont settled in Paris where before long he would marry.

To return to Australia, Loir had convinced Pasteur to let him proceed with the anthrax vaccination program in New South Wales on a user-pays basis, which could make the Pasteur Institute a pretty penny; and to also continue the pleuro-pneumonia research for the Queensland Government. And he had not lost sight of their chicken-cholera bid for the rabbit prize – whilst working on other projects in Australia, Loir, in his own words, would be 'waiting for public opinion to demand that this scientific, as well as practical method, be put into practice'.[343]

As soon as Loir arrived back in Sydney he took up residence on Rodd Island. The previous staff had been let go, so Loir found his own personnel. As cook, he employed a French-speaking coloured native of Martinique who had served the governor of New Caledonia, Admiral Amédée Courbet, and who'd later run a French restaurant in Noumea. As soon as the cook had the job, he went out and married a fair-haired Irish woman who became Loir's housemaid. 'I therefore found myself with good service and excellent cooking,' Loir was to write. As his handyman, Loir hired a powerful islander from Fiji.[344]

Almost certainly, Loir had left the suitcases of anthrax vaccine that Pasteur had sent him the previous year in cool storage in Sydney, quite probably in the care of Government Analyst Hamlet at the Government Laboratory. Retrieving some of the stored glass tubes, in the first week of July Loir took the train south, heading for Thomas Hammond's property just outside Junee where the anthrax trial had taken place twenty months before. Loir had arranged to meet Arthur Devlin there. Devlin at that time was reeling from the loss of Uarah Station: only months earlier, the Scottish Australian Investment Company had foreclosed on Devlin's mortgage and he'd been forced to turn his back on the property he had laboured over for so many years. His new young friend Adrien Loir had thrown Devlin a lifeline; Loir now employed Devlin to supervise all anthrax vaccinations in New South Wales for the newly constituted Pasteur Institute of Australia, of which twenty-seven-year-old Loir was officially Director.

There at the Hammond property, Loir taught Devlin how to vaccinate stock with the two-dose anthrax vaccine, then stood back and watched while the Australian personally vaccinated one hundred sheep provided by Tom Hammond, an enthusiastic supporter of anthrax vaccination. Devlin then set off to sell the idea of vaccination to the sheep graziers of New South Wales, and to train teams of vaccinators throughout the colony.

The anthrax vaccination program immediately met opposition from anti-vaccinists, led by none other than Dr Oscar Katz, the Rabbit Commission's former 'expert'. Katz had returned to Sydney looking for work after spending some months in Melbourne, having been employed by Professor Harry Allen, President of the Rabbit Commission, to carry out experiments for the Victorian Government's Sanitary Commission inquiry of which Allen was also the head. 'It was not to be expected that the introduction of this system of "vaccination" would be allowed without some opposition,' said the *Australasian Pastoralists Review* of 15 August. 'Dr Katz entered the lists as its warmest enemy.' According to the *Review*, Katz contended that a mortality rate of perhaps 200,000 head a year from anthrax, out of a total sheep population of fifty million, was not worth the risks attached to vaccination.

What Katz either did not know, or did not acknowledge, was that anthrax in the colony tended to be localised, killing large numbers of sheep on several hundred properties while leaving others unscathed. As Arthur Devlin could so painfully testify, if anthrax hit, it could wipe out thousands of sheep on a single property and destroy its viability. Katz was saying that it was more economical to simply let 200,000 sheep a year die from anthrax. Yet Katz could not guarantee that the following year the death rate would not multiply, or that the disease would not spread.

And, as Pasteur had learned, once an animal dies from anthrax, unless the animal is swiftly burned – as was the practice in France – the bacillus leaches into the soil where it can remain virulent for up to twelve years. In the past, Australian farmers had simply let stock that died from anthrax rot where they fell, thereby impregnating the soil with anthrax and ensuring its recurrence in stock that later grazed on that land. This, Adrien Loir had discovered, was why the lethality rate of anthrax was so much higher in Australia than in France. Now, as part of the Pasteur Institute of Australia's vaccination program, Loir urged stockowners to swiftly burn all anthrax victims. Devlin, meanwhile, was the perfect man for the job of anthrax vaccine salesman – he was able to warn his colleagues that unless vaccination was carried out in infected areas many landowners in those areas could be ruined by fresh outbreaks, just as he had been.

For the sake of protecting their flocks and their profits, many squatters in the infected areas were prepared to listen to Devlin, and to ignore Katz and other anti-vaccinists who cautioned that vaccination only caused a variety of other diseases. Loir and Devlin were also able to remind stockowners that anthrax vaccinations had been safely conducted in France for years, and to tell them that in France insurance companies would not insure stock unless they had received the Pasteur anthrax vaccine.

Between July and November that year, Arthur Devlin and his teams would vaccinate close to 200,000 sheep with Pasteur anthrax vaccine, with farmers paying three pence a head for the vaccination of their stock. Overheads absorbed twenty-five percent of the vaccination income, but by the end of that year Loir was still able to transmit somewhere in the vicinity of 450,000 francs to Pasteur in Paris – almost as much as the rabbit prize was worth. Had Pasteur permitted Loir to launch the anthrax vaccination program in 1888, he would have had that money a year earlier, and would have slept far more easily in the lead-up to the opening of the Pasteur Institute.

Now that Loir was generating such large profits for Pasteur – having contributed some 700,000 francs to the Pasteur Institute's funds via the pleuro-pneumonia cheque and the first year's anthrax vaccine sales – there could be no argument about the value of his remaining in Australia. And this suited Loir admirably. It had not taken him long to become fond of this vast, sunburnt country. And he embraced the laconic country people who were devoid of the pretensions of city socialites and degree-conscious academics. These were tough, grounded people who took a man as they found him, and stuck by him through thick and thin – 'mateship', these Australians called it.

Loir had acquired a camera to photograph laboratory slides, so whenever he travelled through the interior of the country he took it along and photographed the countryside and its inhabitants. He became friendly with Aboriginal people, no doubt via Aboriginal stockmen he met on various stations. Fascinated with their culture, he collected indigenous weapons, photographed tribespeople in their ceremonial markings, and even attended a tribal corroboree dance. He also made plenty of friends in Sydney, male and female, especially among up-and-coming professionals. It became his habit to invite selected friends to Sunday lunch on Rodd Island where he would wow them with the stunning island location and mouth-watering food from the best French chef in New South Wales.

Just six months after arriving back in Sydney, in January 1891 Loir heard that the most famous actress in the world, France's Sarah Bernhardt, was coming to perform in Sydney in the middle of the year as part of a world tour. According to the actress's reputation, like a rock star today she made extraordinary demands of her producers, travelled with a vast entourage, surrounded herself with a menagerie of pets, sometimes dressed as a man, and slept in a coffin. As Adrien Loir was to discover, only the story about the coffin was incorrect.

One London critic said of 'the Divine Sarah' at the time: 'For all her eccentricities, it must not be forgotten that she possesses enormous talent, and hardly falls short of being a woman of genius.'[345] Said the *Illustrated Sydney News* as her arrival in Sydney approached: 'Critics have denounced Sarah Bernhardt as "nothing more than an eccentric personality", while others have applauded her as "the embodiment of the subtle arts". We think the truth may lie somewhere between.'[346] Very feminine, very sexy, and very famous in America and Britain as well as in Europe, Sarah counted among her more passionate admirers the Prince of Wales, Mark Twain, D.H. Lawrence, Victor Hugo, and the Czar of Russia. Psychologist Sigmund Freud kept a picture of her on his wall. And her only son was sired by a Belgian prince.

As Madame Sarah's arrival in Australia grew closer, excitement mounted in Sydney, fed by a constant stream of articles about her in the press. An official welcoming committee was formed, headed by William MacMillan, Treasurer in the latest Parkes Government. Adrien Loir was one of half-a-dozen locally resident French citizens co-opted onto the committee, with Loir, at just twenty-eight, by far its youngest member. His fellow French committee members included fifty-year-old musician Henri Kowalski, who styled himself 'the prince of the pianoforte' and presided over the Sydney Choral and Orchestral Society; Kowalski was considered notable enough for Tom Roberts to paint his portrait a few years later. Another Frenchman on the committee was sixty-two-year-old Horace Poussard, a violinist and composer who made his primary living as a music teacher in Sydney.

At 10.00 on the sunny morning of Tuesday 26 May, Loir was one of the two hundred official guests crammed onto the steam-launch *Daphne* when it left Circular Quay to the sounds of the Centennial Band playing the *Marseillaise*. The plan was for the launch, which was decked out with red, white and blue bunting and French tricolour flags, to meet the Royal Mail Steamer *Monowai* as it arrived from San Francisco via Honolulu and Auckland, bringing Sarah Bernhardt and her acting troupe to Australia's shores. Problem was, the RMS *Monowai* had actually arrived early, in the middle of the night, so the *Daphne* merely trundled around to Darling Harbour where the steamer had tied up at the Margaret Street Wharf. There, on the steamer's upper deck, stood the Divine Sarah herself, queen-like, waving modestly. Cheer after cheer went up from those aboard the *Daphne.*

The guests transferred to the wharf and then boarded the *Monawai*, trooping with anticipation into the steamer's saloon. There Madame Bernhardt stood waiting for them, surrounded by massive arrays of fresh flowers, and smiling faintly. Over the years she had suffered through a thousand receptions such as this; now, with practised patience and grace, she greeted every member of the welcoming party, offering a restrained hand to each one. Frenchman Henri Kowalski had been boasting in Sydney that he knew Madame Sarah personally, and he now attempted to remind her of the occasion on which they had met: 'Long ago, at the Boulevard Malsherbes, where we danced a polka together,' he said. Madame Sarah could not remember the occasion, but this did not prevent Kowalski from acting as though he were her long-lost friend, introducing other guests to her as they filed past.[347]

Adrien Loir took his turn in the reception line. Madame Sarah glanced at him briefly as he shook her hand but said nothing, just nodded. She was all that her advance publicity had painted her to be: quite tall, meaty, with a long nose, a pert mouth and dreamy eyes. Her hair was red and frizzy, and not particularly long; she frequently wore wigs on stage. Her skin was soft and luminous. She looked to be in her thirties yet Madame Sarah was in fact forty-six. Nevertheless, she would continue to convincingly play the roles of young women on stage for decades to come, in a career that still had thirty years to run and that would see her end up as an early star of silent movies.

Sir Henry Parkes was not among those who had come to greet Madame Sarah. In fact, he was not to meet her at all during her extended visit. Sir Henry did not go to the theatre, was not a member of a single gentleman's club. His idea of socialising was a cosy dinner party at home surrounded by good friends. Although he generally abstained from cigarettes and alcohol, he shared one thing in common with Madame Sarah: a taste for the occasional glass of champagne. To welcome Sarah Bernhardt on behalf of his Government, Sir Henry sent a loyal friend and cabinet minister, the Postmaster-General, Daniel O'Connor.

A tall Irishman from Tipperary, O'Connor had thick white hair and a flowing white beard in the Parkes mould, which made him look much older than his forty-six years. O'Connor would be lampooned in the colonial press for his long welcome speech to the Divine Sarah, which his broad Irish accent and attempts at French sometimes rendered incomprehensible and at other times hilarious. When he referred to France as 'Ley bellee Frangshay', even his audience aboard the *Monawai* roared with laughter. When the time came to respond, Sarah merely said to the guests, in heavily accented English, 'I thank you very much for your great kindness to me.'[348] She was then whisked away aboard the *Daphne* to Circular Quay where an excited crowd thousands strong awaited her. From the launch, it was a red-carpeted walk through the adoring crowd to a waiting coach, and then it was off to the Grosvenor Hotel.

That night, Adrien Loir read in one of the evening papers that Madame Sarah was deeply upset that, because of the colonies' strict anti-rabies quarantine regulations, her two pet dogs had been taken from her by the authorities and sent to Long Island at the mouth of the Hawkesbury River, north of Sydney, to be held in quarantine until she left the continent. Loir had an idea. Bright and early the following morning, he visited the Grosvenor Hotel and asked to see Madame Sarah, saying that he had good news for her. He had to wait some time: Sarah did not rise before 10.00. But finally he was ushered into her suite. Loir gushed out his idea, telling the seated Sarah that he was certain he could convince the local authorities to allow him to care for her dogs on Rodd Island. 'She gladly accepted my offer, thinking her pets would be in good hands with me,' Loir was to say.[349]

Loir hurried away to keep his word to the Divine Sarah, knowing that she was due to leave Sydney a little before 5.00 that evening for Melbourne aboard a specially hired train. But he found that his friend, Government Veterinarian Edward Stanley, didn't like his idea. So he then tried his best friend in the colony, Chief Stock Inspector Tiger Bruce, with whom Loir had recently written an article about stock diseases in New South Wales which, to Bruce's joy, had been published in the *Annals of the Pasteur Institute* in Paris. But Bruce wasn't encouraging either. At 2.30, Loir, feeling that he had let his countrywoman down, attended a civic reception for Sarah at Sydney Town Hall which was crowded with five hundred invited ladies and gentlemen from the 'hupper suckles' – as the local satirical magazines called Sydney society's upper circles.[350]

There was white-bearded Daniel O'Connor again, glass of champagne in hand, talking animatedly to Sarah. Loir pushed through the throng and slipped in beside Madame Sarah just as O'Connor was saying, 'Madame, the artistic capital of the world is no longer Paris, since Sydney is honoured with your visit.'

Loir leant forward and whispered to Sarah in French: 'A minister is talking to you. Ask him to allow your dogs to be placed in my care.' She glanced around at Loir, and then told him to put the proposal to O'Connor himself as her English was poor. So Loir put the question to the Postmaster-General, asking on Sarah's behalf that her dogs be left in his care on Rodd Island.

O'Connor smiled broadly. 'Of course, Madame, your dogs will be placed in the care of your fellow-countryman. It is the least we can do for you.' After Sarah thanked O'Connor and then turned to talk with someone else, the Postmaster-General leaned close to Loir and said in a low voice, 'That will not be possible.'

Loir, surprised by this sudden *volte-face*, studied the Irishman for a moment, then said determinedly, 'I will prove you wrong, Minister.' When O'Connor looked at him questioningly, Loir explained that by the stroke of a pen the government could turn Rodd Island into a quarantine annex, which would permit Loir to keep Madame Bernhardt's dogs on the island with him – officially in a state of quarantine.[351]

O'Connor beamed. Jumping at the idea, the Irishman took Loir aside and wrote a letter to his cabinet colleague, the young Mines and Agriculture Minister Sydney Smith. Loir then hurried away with O'Connor's letter in hand. He managed to catch Minister Smith on the front steps of his ministry as he was heading off to the Town Hall reception in his best grey frock-coat and top-hat, a flower in his buttonhole.

Gregarious thirty-five-year-old Smith, a short man with a handsome face that not even his shaggy beard could disguise, greeted the young Frenchman cheerily. 'Loir, my dear fellow. How's it going with your vaccinations?'[352]

Loir told the minister that he had come on another vital matter, and handed over Daniel O'Connor's missive. There on the steps Smith read the letter. Although sceptical of the proposal, the minister then took Loir back up to his office and sent for Tiger Bruce. When the Chief Inspector of Stock joined them, he supported Loir's idea. Just the same, that night, when Sarah's train steamed out of Redfern Terminus, heading south for Melbourne in the winter dark, the Divine One's pampered pooches were still imprisoned behind wire on Long Island.

But four days later, when Sir Henry Parkes and his cabinet sat down with the Governor at the next meeting of the Executive Council at 121 Macquarie Street, Loir's proposal was put up by Sydney Smith and approved by the Premier and his cabinet. Rodd

Island was officially declared a quarantine annex, and Sarah's two dogs were brought back from Long Island and given over to the custody of Adrien Loir on Rodd Island. One of the dogs was a tiny pedigree female toy terrier named Chouette – Sarah had paid 6000 francs for her, outbidding the Baroness de Rothschild. The second dog, Star, a sky-terrier, was not much larger than 4600-gram Chouette.[353] By making a home for Chouette and Star, Loir had just given himself an entrée into the world of the Divine Sarah. And his life was about to change forever.

25

THE SARAH BERNHARDT AFFAIR

FOR FOUR WEEKS, Sarah Bernhardt performed in a sell-out season in Melbourne followed by a triumphant week on the Adelaide stage.

While Madame Sarah was away from Sydney, Adrien Loir worked diligently at his Rodd Island laboratory. He made new anthrax vaccine for Arthur Devlin's latest winter round of vaccinations. He carried out anthrax research commissioned by the New South Wales Government involving Australian marsupials. He worked on the continuing pleuro-pneumonia research for the Queensland Government. And, as he had every day since his return to Australia, he scoured the press for items referring to the unabated rabbit plague that might give him ammunition for a renewed chicken-cholera campaign. He cut out every press report of interest and pasted them into scrapbooks that he assiduously maintained while he was in Australia. All through this time he also took doting care of little Chouette and Star.

At 2.27 pm on Tuesday 7 July, Sarah Bernhardt's special train from Melbourne steamed into Redfern Terminus bringing the star, her acting troupe and their baggage back to New South Wales for the tour's Sydney season. The colourful French troupe consisted of six actresses, twenty-one actors, a producer, twelve stage hands, dressers and personal servants. The baggage included stage scenery, props and costumes for a number of productions: Madame Sarah performed a different play every night on tour. Sarah's costumes alone occupied eighty of the company's one hundred and thirty-seven trunks, with each trunk a metre and a half high.

The costumes, along with the staging, were half the reason Sarah so dazzled her audiences. Her costumes were worth a king's ransom – Sarah had brought 550 dresses, 250 pairs of shoes, and numerous coats, cloaks, hats and wigs to Australia, with every item specially handmade for her in France.[354] One of the dresses she wore in her most famous role, that of Marguerite Gautier in *La Dame aux Camelias*, or *Camille* as it was known by English-speaking audiences, was adorned with real pearls; on an earlier tour of the US, the dress had been valued at $10,000 – $4 million in today's money.[355] And then there were the essential supplies – Sarah's party had landed in Sydney in May with eighty-one cases of champagne, two kegs of brandy and a keg of whiskey.[356]

Leaving the train, Madame Sarah was driven in a hired coach fit for royalty to the brand new Hotel Australia on the corner of Castlereagh and Rowe streets. Many locals had doubted the viability of this narrow, seven-storey tower of luxury in the heart of the business district when Sir Henry Parkes had laid its foundation-stone a little over twelve months earlier. Trading on the fact that 'Sarah Bernhardt slept here', the Hotel Australia would become Sydney's most elite hotel, attracting a host of famous guests before being demolished in the high-rise rush of the 1970s. Final fit-out was still in process and the official opening still two weeks away when, on 7 July, the Divine Sarah glided by the Australia's polished granite pillars and checked in as the hotel's first guest, before taking the brand new American hydraulic steel-ram elevator to her floor.

Sarah's eight-room suite, occupying the entire front of the hotel overlooking Castlereagh Street, comprised a sitting room, a dining room, a master bedroom for Sarah and another for her companion of some years, Suzanne Seylor, and four smaller rooms for servants. All the rooms had high ceilings, fireplaces and thick Axminster carpets, while the sitting room also featured imported antique furniture, a piano, and massive indoor plants that gave the room a touch of the exotic.

It was here in her suite at the Hotel Australia that Adrien Loir found Sarah that same afternoon. Feeling pleased with himself, he told the Divine One that he had arranged for Chouette and Star to join him on Rodd Island, and handed her a permit from the government allowing her to land on the island, which was now legally off-limits to unauthorised visitors. This meant that Sarah could see her two little dogs any time she wished while avoiding crowds of fans. Loir now boldly invited Sarah to visit the island two days later, on Thursday, to see her dogs, telling her he would even provide lunch cooked by his excellent French chef. When Sarah at once agreed, Loir pushed his luck and asked if Madame would mind if he invited a few of his Sydney friends to join them for lunch.

Thoughtfully, Sarah studied the young man for a moment, then said, 'I dislike the idea. But if it can be useful to you in any way, do it.' And then she added that as his reward for looking after Chouette and Star, he was to join her for dinner every night after each of her Sydney performances.[357]

Loir looked at the Divine Sarah agog. Months before, he had purchased a season ticket, planning to see every one of Sarah's Sydney performances. But to also join her for dinner afterwards, every night – that was unbelievable! He had heard that she always dined at midnight. How would he return to his island home in the early hours of each morning? He would have to worry about that when the time came. For the next five weeks, Adrien Loir was to lead a life unlike any he had ever dreamed of.

On the next night, a Wednesday, Loir was in the vast crowd that pushed expectantly into Her Majesty's Theatre at the Haymarket as rain poured down outside. As he entered the theatre, Loir, like all the other rain-soaked theatregoers, was handed a playscript printed in English. Before the Bernhardt troupe arrived in Australia, Loir had assumed that in an English-speaking land Bernhardt's plays would be performed in English. But no, every word would be spoken in French. To help the audience follow the story, the impresarios who had brought Madame Sarah to Australia – James Cassius 'J.C.'

Williamson, an American actor born in Pennsylvania and raised in rural Wisconsin who had become Australia's top theatrical producer, and Sarah's American agent Henry Abbey, a former director of New York's Metropolitan Opera – had each play's script translated into English and distributed to audience members. When Sarah had learned of their plan to keep the house lights burning throughout every performance to enable the audience to follow their scripts, Sarah had demanded that all the light-shades be painted pink, to soften her appearance. Rather than incur this expense, as a concession to the Divine One the management dimmed the lights a little when the curtain rose.[358]

Sarah's first Sydney production was Alexander Dumas the Younger's famous *Camille*. The play had been performed in Sydney several times before, but not with the Divine Sarah in the lead. Critics considered the play much too sentimental, but they loved Sarah, and so did the audience. As always, her death scene mesmerised everyone in the theatre. After the many curtain calls, Adrien Loir made his way backstage and joined the throng of well-wishers that flocked into Sarah's long, narrow dressing room. Once Sarah had changed, she invited Loir to join Suzanne Seylor and herself in her coach for the ride back to their hotel. To Loir's astonishment, it was going to be just the three of them who would dine in Sarah's suite that night.

Suzanne Seylor, only now in her thirties, had run away from school to join Sarah's company. At first she had served as Sarah's maid, but Sarah had spotted acting talent and had put Seylor on stage. The petite Seylor had a boyish face and sometimes played page boys. On other occasions she took musical roles, for Seylor had the sweetest singing voice in the entire Bernhardt company. In addition, Seylor served as Sarah's personal assistant – where Sarah went, so too did Seylor. Realising long before Loir did that Madame Sarah had taken quite a shine to the charming if unworldly young French scientist, Seylor would soon take a sisterly interest in him.

It was past midnight by the time the trio returned to the Hotel Australia. In the Divine One's suite, Loir was left sitting in front of the fire with Seylor while Sarah disappeared into another room.

As she did every night after a performance, Madame Sarah took a bath. When she returned, it was in a beautiful negligee of long, grey, flowing French silk. The hotel's chefs had long before gone home. But before they did, they had prepared a cold meat supper for Sarah, to her precise instructions. Also on Sarah's orders, the suite's dining room had been equipped with a small gas stove. Calling Loir to assist her, Sarah cooked a meal of eggs for the three of them in a silver pan. Over the next five weeks, sometimes it would be fried eggs but most nights, like tonight, it was scrambled eggs, to Sarah's own recipe. With Loir standing at her side, she added a spoonful of beer to the eggs as she beat them, and then sometimes a little ham. 'I assisted her by handing her the salt and pepper,' Loir would later say with a hint of amusement.[359]

Loir sat with Sarah and Seylor, supping and telling them about his great-grandfather Joseph Loir who'd been a veterinarian in Napoleon's army in Egypt; Napoleon had given him one of his hats and ordered him to marry the daughter of the army's chief veterinarian. Loir's great-grandfather had gone home to France with the emperor's hat and a new bride.[360] Loir told Sarah about his years with Louis Pasteur, and of all the frustrations he had experienced in Australia with the Rabbit Commission, and its German doctor and German-favouring commissioners. Five years later, in an interview with *Le Figaro* in Paris, Sarah would tell the paper that the French community in Australia was dominated by the Germans; this belief seems to have stemmed from her conversation with Loir.[361]

Sarah, an intensely private person, would have told Loir little about herself. It was common knowledge that she was of German-Jewish descent, and that her mother, a courtesan, or mistress of wealthy aristocrats, had sent little Sarah to a Catholic nunnery where she was raised. Sarah best liked to talk about the outrageous things she did – she loved to shock – and about her animals; her collection back home in France included a cheetah, chameleons and snakes. Even here in the Hotel Australia suite, Sarah, Seylor and Loir were not alone: since arriving in the Antipodes, Sarah had been given a St Bernard dog which she'd named Auckland after the first Australasian port she had touched; she'd also purchased a range of animals which all lived in the suite with her. Sarah's favourites were two Victorian possums, one of which she named St Kilda. While Loir dined, he was conscious of the pair of possums clinging to the suite's plush curtains and watching his every mouthful.

And then the evening was over. As 2.00 am approached, Sarah rose, kissed Seylor on the cheek and went to bed. Loir does not say where he spent that first night, but it is likely that he bedded down in one of the two vacant servants' rooms, for Sarah seems to have brought only two maids with her to Australia.

Next morning, late, Loir slipped out of the hotel and hurried to Circular Quay; this was the day that he'd arranged for Sarah's first visit to Rodd Island. After Sarah had allowed him to invite a few friends to join them on the island for lunch he had organised a party of twenty-three: 'I approached a friend, a young lady of Sydney's high society who wanted to meet Sarah Bernhardt,' he would write, 'and asked her to organise a gathering of approximately twenty ladies.' At 11.00 am, as Loir hurried onto the wharf, he was pleased to see 'a group of pretty, elegantly dressed ladies' waiting for him.[362]

Loir had also invited the chief medical officer of the colony's little army to join the party, along with another local doctor whom Loir had befriended. As Loir arrived at the wharf, the army doctor came toward him, smiling broadly. 'Loir, how did you manage to assemble all these ladies?' he asked. 'They are Sydney's most elegant.'[363]

Loir had arranged for a small government steam-launch to take the party to his island. After Sarah and Seylor arrived in Sarah's carriage, everyone boarded the launch and it set off for the pleasant, thirty-minute trip around to Iron Cove. To Loir's surprise, most of the well-educated young ladies could speak a little French, and Sarah chatted happily with them as the launch traversed the harbour. When the boat pulled into the Rodd Island jetty and the gentlemen helped the ladies disembark, Loir offered Sarah his arm and led her up the little road to the buildings on the peak. Looking around with delight, Sarah said to Loir, 'How poor an artist you are! Had I been given this island, what good use I would have made of it, with cannons placed here and there! I would have turned it into a small fortress. On my return to France, I want to settle on an island.'[364]

The lunch provided by Loir's French chef was a triumph, making the day unforgettable for all concerned. When Loir gave his guests a guided tour of the island and its facilities, Sarah spotted the two vacant bedrooms in the residency; later, as the party was boarding the launch to return to the city, she said to Loir, 'There will be no performances on Sunday in this English country. I wish to come here every Sunday, accompanied by Mademoiselle Seylor, for twenty-four hours, so that I may be by myself and rest.' Loir immediately agreed. 'But you must have no other guest,' Sarah added. Again Loir enthusiastically agreed.[365]

Once again that night, Loir dined with Sarah at the Hotel Australia, this time after she had performed in Victorien Sardou's play, *La Tosca*. Nine years later, Italian composer Puccini would turn Sardou's popular French play into his even more famous opera, *Tosca*. Again, apart from Seylor, Loir was Sarah's only dinner guest. Throughout her Sydney stay, Sarah would receive numerous invitations to attend dinners and soirées of one sort and another, and several people would later claim that she had dined with them; but in reality Sarah reserved all her Sydney dinners for just Adrien Loir, Seylor and herself.

Several times Sarah hosted large luncheon parties in her hotel suite, and on 30 July she attended an afternoon tea held by the very social Sarah Jenny Fischer at her Potts Point mansion. Mrs Fischer was the estranged wife of the same Dr Fischer who had brought the microbes to Sydney from Robert Koch in Berlin as a part of Camac Wilkinson's attempt to thwart Louis Pasteur; but Sarah would of course have been unaware of this connection. Loir never accompanied Sarah on these very public social occasions; she kept her time with him hidden from general view.

Loir was very quickly accepted by Sarah's entourage who realised that he was their boss's latest plaything; they knew that Madame's preference was for younger men. One of Sarah's former lovers, Edouard Angelo, was one of the two young actors in the company on this world tour; most of the male French players were middle-aged. The other youth, the tall, curly-haired twenty-something, Albert Darmont, had spent a lot of time in Sarah's company earlier in the tour, working with her on his own play *Pauline Blanchard*: Sarah would give the play its world premiere in Sydney on 22 July. But it is unclear whether there had been any romantic attachment between Darmont and Sarah. If there had been, this had evaporated by the time they reached Australia, for Darmont is only mentioned twice in Sarah's company offstage, and each time it was with a number of others.

With Loir's overnight status as Sarah Bernhardt's new close friend, he was allowed unrestricted access to the theatre and often would watch from the wings as Sarah performed. On Friday 10 July, just four days into the Sydney season, Sarah was due to perform *Fedora*, a play especially written for her by Victorien Sardou. She chose plays containing death scenes that she could milk like no one else; or scenes that, at the very least, allowed her to melodramatically swoon and faint. *Fedora* was no exception: at the beginning of the play, Sarah's character finds her lover Vladimir dead in bed; she swoons and collapses on top of him. Over the years, many men had vied to play the brief, non-speaking part of Vladimir; even the Prince of Wales had scandalously filled the role opposite Sarah on the London stage.

On the night of 10 July, Loir was standing in the wings watching final preparations on stage just before the curtain went up on *Fedora*. Duberry, the actor who was to play the part of Vladimir, turned to Loir and said, 'Would you like to play the dead man?'[366]

All the other players thought this an excellent idea and the beautiful Jeanne Mea, the company's second principal actress after Sarah, took Loir's hand and led him to her dressing room. Mea, in her thirties, was the only daughter of playwright Alexander Dumas the Younger and Nadeja Varyschkine. In the words of one adoring Australian journalist, Mea was 'a tall and radiant nymph with henna-dyed hair and a big emerald ring on her finger'.[367] Loir called her 'an actress of imperial bearing'.[368] Using lipstick, Mea simulated a blood-stain on Loir's shirt, then sent him back to the stage.

But when Sarah's French producer Louis de Cosse-Brissac saw Loir in position on the bed, he demanded to know, 'Who is on the bed?'

'It is Loir,' said Suzanne Seylor. 'You must let him stay there. He is having so much fun!'

Cosse-Brissac frowned. 'But, does Madame Sarah know?'[369]

The thought that he might annoy Sarah was enough to terrify Loir, who jumped up and departed the stage, leaving Duberry to hurriedly take his place as Vladimir just as the curtain was about to rise.

That night, when Loir had supper with Sarah and Seylor as usual, Seylor said, 'Poor Loir was rather disappointed. He wanted to play the part of the dead man but was not allowed to do it.'

Sarah scowled. 'Why not?'

'Cosse-Brissac prevented him.'

'What a fool!' Sarah raged. 'It is none of his business.' As Loir came to learn, Sarah was on bad terms with her producer. Sarah now made a snap decision. She had yet to announce what play to put on for her final Saturday matinee; now she said, 'It shall be *Fedora*. And this time, Loir, you will play the dead man.'[370]

Two days later, on her first Sunday in Sydney, Sarah slipped away from the city on the government launch with Loir and Seylor, and they chugged around to Rodd Island where government signs on all four corners of the island warned the public to keep off because this was a quarantine site. The two women spent the night at the residency with Loir, each with a bedroom to themselves. Just before 8.00 am on the Monday, Loir arose and went duck-hunting from the island's shore. The boom of his shotgun awoke Madame Sarah; she was not pleased. The victim of her reproach, Loir learned to never again make a sound before 10.00 am so that he did not wake the Divine One and ruin her beauty sleep.

All the following week, Sarah performed one play after another to packed houses. Every night, Loir watched from the wings then dined on eggs and ham with Sarah and Seylor. On Sunday the 19th, while the remainder of the Bernhardt company went off on excursions of their own, Sarah and Seylor again joined Loir on Rodd Island. This time, Sarah also brought along Jeanne Mea. The actress who was due to play the part of Louise in the play *Frou-Frou* on the Monday night had taken ill so Sarah had asked Mea to fill in for her. Mea knew the part but was nervous about playing it opposite Sarah. So, for more than an hour that Sunday morning, while Loir lounged on his bed and watched in admiration, Sarah sat with Mea on a small settee in

Loir's bedroom and gave her 'a carefully nuanced reading-rehearsal'. The privileged Loir was entranced. 'What a teacher!' Come Monday evening's performance he said, 'The dialogue between Sarah and Mea was wonderfully true to life.'[371]

That same Sunday, Loir's moustache became the focus of attention. Madame Sarah had a fixation about facial hair; she didn't like the male fad for beards and moustaches. While she tolerated it in her actors, she expected them to keep their facial hair neat and to always come to performances freshly shaved, fining them, even fining the extras, if they didn't shave. (One of those extras during the Sydney season was a short young Labor Party politician who was to become Australia's Prime Minister during World War I, Billy Hughes.) Over the last day or two, Sarah had been urging Loir to shave off his moustache, of which he was very proud. When they rose from lunch, Sarah said to him disapprovingly, 'You still have a moustache.'

As Sarah prepared to go fishing in one of three small boats that Loir kept at the island, Jeanne Mea took Loir aside. 'You know what remains to be done,' she said. 'Come to your room and I will help you shave.'

Minutes later, Mea led Loir down to the island's jetty. Sarah was in a dinghy, equipped with fishing rod and line. With Loir's big, brown Fijian handyman dragging on the oars, the boat was pulling away from the jetty and out onto the cove.

'Madame Sarah!' Mea called, then pointed to Loir's top lip which was now devoid of its moustache.

Sarah beamed, then shouted, 'I shall kiss him on my return.'[372]

Loir says that Sarah kept her word, and indeed did kiss him when she returned from her afternoon's fishing on the cove.[373] He would go into no more detail about their relationship; this was as intimate as he would become in his revelations. Many years later, in France, it would be said of Loir, 'He *had* an actress on an island in Sydney harbour.' And it became accepted fact that the bedded actress in question was Sarah Bernhardt.[374] It seems that the intimate physical relationship between Bernhardt and Loir began there on the island that Sunday 19 July, with Seylor occupying one of the residency's three beds that night, Mea the second, and Loir and Sarah the third.

It is likely that Loir shared Sarah's bed at the Hotel Australia six nights a week thereafter, while she shared his bed at Rodd Island on Sunday nights. So fond of Loir did Sarah become that rather than leave Sydney, and leave him, when her scheduled four-week season came to a close and she was supposed to travel up to Brisbane to do a week's run there, she abruptly cancelled the Brisbane season.

Brisbanites never knew that they had Adrien Loir to blame for Divine Sarah's failure to play in their city. The excuse for the Brisbane cancellation was that Madame Bernhardt was too tired; yet she wasn't too tired to do an extended season in Sydney: the company transferred to the newer but smaller Theatre Royal in Castlereagh Street for another six evening performances and the final matinee – *Fedora*, with Adrien Loir playing the dead Vladimir.

After his stage debut that Saturday afternoon, Loir dashed along Castlereagh Street to the Hotel Australia in his theatrical costume, complete with fake blood, to show Suzanne Seylor, who had been instrumental in securing him the part – Seylor, who had no role in *Fedora*, had remained back at the hotel. Loir went proudly to her room and showed off his costume. But when he went to unbutton his waistcoat to remove his bloodied shirt-front, Seylor suddenly went pale, thinking that Loir wanted to add her to his short list of conquests. Jumping up, she dragged on the bell-pull, summoning the maid. When Loir laughed and she realised her mistake, Seylor covered her embarrassment by asking the maid to serve them tea.

That night, for her final performance in Sydney, Sarah did a reprise of *Camille*. For the last bows, which seemed to go on for hours to the accompaniment of cheers, applause and stamping feet from the ecstatic packed house, together with several renditions of the *Marseillaise* and a truckload of flowers laid at the Divine One's feet, Sarah was joined on stage by all the players, impresario J.C. Williamson, and the hangers-on who'd attached themselves to her during her Sydney stay. Those hangers-on retired with Sarah to her dressing room shortly after where the star, still in the white muslin of the last act, in an atmosphere of 'kitten French and gushing huggery' according to female columnist Sapho Smith, accepted their devotions.[375] And then it was all over – for the public, at least. Sarah had another final act to perform, with Adrien Loir.

On the Divine Sarah's last Sunday in Sydney, she took all her actors and actresses to Rodd Island for lunch and dinner with Loir. But she had laid down ground rules for Loir: 'I will come to your place with my company – twenty-seven people – only on the condition that all expenses will be paid for by me. I will bring provisions for both meals.' Then, once the party arrived on the island, and as the first champagne corks were popping, Sarah asked Loir for the keys to the laboratory and stables. She then clapped her hands and called the actors and actresses together. 'My dears, you can go anywhere except into the stables and the laboratory.' Besides, they were locked she said, holding up the keys. 'Now, let us play hide and seek. All the ladies come with me. The men will look for us on our signal.'[376]

Loir and the other men waited; then, on hearing Sarah's distant signal, went in search of the ladies who had trotted away, champagne glasses in hand. Despite scouring the island, nowhere could the men find the women. It took girlish giggles above them to betray the ladies' hiding place: Loir looked up and there were Sarah and her five female companions on the roof of the residency. They had climbed up a ladder which they had then pushed away. It was the start of a memorable day. That evening after dinner, the very merry company steamed back to the city, leaving Sarah and Seylor to spend their last night with Loir.

Lunch the next day was a sombre affair for this was to be their last time together in Sydney. As they ate, Sarah said to Loir, 'What are you doing here?' By 'here' she meant Australia, this remote southern land.

Loir, defensive of his lifestyle and of his affection for Australia, responded that Sarah had witnessed his way of life for the past month so, he said, she already knew the answer to that question.

But Sarah wasn't talking about Loir's work. 'There are women in France,' she said.[377] Sarah wanted Loir to go home and marry a French girl.

Over the previous weeks, Loir may have revealed to Sarah that he had a ladylove in Sydney – possibly the young socialite who had organised the party of local beauties to visit Rodd Island with Sarah in July. Loir may have also revealed that this young lady, an Anglican, had been pressuring him to consider marriage; he may have even promised to marry her. The clues pointing to these possibilities lie in one of Loir's scrapbooks for this period. Twice he had pasted in classified newspaper advertisements from authorised marriage celebrants who offered fast weddings out of church. Like his parents, Loir was a Roman Catholic, though then not a particularly active one. He also pasted into his scrapbook several press stories about a socialite who took a gentleman to court, sued him for breach of promise, and won. These personal items stood out in Loir's vast press clipping collection: most of the other items, apart from his rural photographs, invitations, and a pressed flower or two, were all about rabbits, chicken-cholera, sheep, anthrax, pleuro-pneumonia, Aborigines, or himself.

It is clear that Loir was loving it in Australia and, after six years as Louis Pasteur's slave in France, was enjoying his independence. Marriage simply did not figure on his list of things to do in the foreseeable future. But Sarah Bernhardt had other ideas for Loir. Not that she wanted him to marry her; Sarah had long ago resolved that marriage was not for her. But she did want to see Loir well-married to a good French girl from sound French stock. Sarah suggested that in a few months, as soon as the current vaccination season was over, Loir should go home and find himself a wife. Loir countered that he had no plans to return to France for at least a year, and that he had a busy slate of experiments ahead of him which would take at least six months. Sarah, determined to get him back to France as quickly as possible, made Loir promise that he would meet her for lunch in Paris in the northern spring, when her world tour would take her back to France for a month.

Late that afternoon, Loir sadly said farewell to Sarah. She did not want him to come to her hotel again, nor to the ship the next day to see her sail for the US. They parted there at his little island paradise, and he sadly watched the steam-launch take Sarah and Seylor down the cove and out of sight.

Next day, Tuesday 11 August, in pouring rain, Loir put little Chouette and Star aboard the government steam-launch and sent them around to Woolloomooloo Bay where the RMS *Mariposa* was preparing to get under way. A crowd of thousands had come to watch Sarah Bernhardt depart. After the final farewells, the official party and favoured reporters hurried ashore, the gangway came in, hawsers were slipped, and the *Mariposa* eased away from Woolloomooloo wharf with a brass band playing the *Marseillaise*.

Onshore, the crowd cheered and waved as Madame Sarah came out onto an open deck, bareheaded in the rain, and stood there alone for a long time, waving back at them. And as the steamer gained speed and moved out into the harbour then swung east, aiming for Sydney's heads and the open sea, the Divine Sarah sadly looked to the west toward Iron Cove and Rodd Island, which lay out of sight, hidden by the city. And with a sigh, she turned for the nearest companionway.

26

DR LOIR AND MRS LOIR

ADRIEN LOIR DECIDED TO take Sarah Bernhardt's advice. He would go back to France as soon as he was able. But not for the reason she had advocated, not to find himself a French bride. Loir decided that it was time to complete his formal education. At his request, Pasteur arranged a study program for him in France that would culminate in his finally taking his doctorate in medicine at the Sorbonne.

But in November, just as Loir completed the latest anthrax vaccination season in New South Wales and was about to sail for home, he was asked by the New South Wales Government to undertake an urgent laboratory investigation. The Sell Brothers Circus, one of America's largest circuses, had just arrived in Sydney from San Francisco, and when its animals were put into quarantine, Loir's friend the Government Veterinarian, Edward Stanley, diagnosed glanders in one of the circus' horses. The American circus owners protested, so the government asked Loir to provide his expert opinion. After testing bacteria from a dead Sell Brothers' pony for ten days, Loir was able to confirm glanders, the first case in the colony's history. The devastated circus owners were refused permission to land any of their horses, which were used not only for haulage but were central to many circus acts.

Finally, just before Christmas, Loir set sail for Marseilles. Thirty-seven days later, at the beginning of February 1892, he reached Paris, and was immediately pounced on by *Le Siècle*: the Paris newspaper had not forgotten about the Australian rabbit prize and Pasteur's claim to it. In an article published on 6 February, the paper reported that Loir had arrived in Paris 'to take some months of rest', and it reminded its readers that Loir had gone to Australia to poison its rabbits but that 'the government refused at the last moment to entrust him with their destruction'. After asking about the operations of the Pasteur Institute in Australia, *Le Siècle* inquired about Loir and Sarah Bernhardt – rumours having apparently reached Paris from members of Bernhardt's touring company, then in the United States, about the pair's relationship. Loir brushed off the question, saying that he had merely looked after Madame Sarah's dogs.[378]

'As for all the rabbits,' *Le Siècle* went on to say after their reporter's conversation with Loir, 'they are constantly spreading, and they have, seemingly, greatly amused the Great Sarah, who nowhere has seen such a large number of rabbits all at once.'[379]

Loir now buried himself in his studies. Officially, he was under the tutelage of former Pasteur pupil Dr Isidore Straus, who was now Professor of Comparative Pathology at the Sorbonne. Part of Loir's time was spent at the veterinary school at Alfort. This was the same veterinary school – the world's first, founded in 1765 – that Pasteur had considered sending Loir to years before. Loir also spent several months living with his parents in his home town of Lyon, to the absolute delight of his mother, while studying at the Lyon veterinary school. Before he left Lyon, he was interviewed by a local newspaper, the *Salut Public*, which was to remark that, with rabbits continuing to ravage Australia, it could not comprehend why the colonial governments would not authorise the use of the Pasteur procedure against them. Loir could only give a Gallic shrug and say that he believed that chicken-cholera was still a viable means of solving Australia's rabbit problem.[380]

While Loir was in Lyon, he had several discussions with Saturnin Arloing, a one-time Pasteur pupil who was now Professor of Experimental Medicine at Lyon University where Loir's father was serving as an honorary professor at the Faculty of Sciences. Knowing that Arloing had conducted research into pleuro-pneumonia and another cattle disease prevalent in Australia, symptomatic anthrax or blackleg, Loir had given the Queensland Government an undertaking that he would try to obtain details of Arloing's blackleg vaccine. The professor generously told Loir how to attenuate his blackleg vaccine, and told him of his research with pleuro-pneumonia: Arloing had never succeeded in cultivating its microbe, but he gave Loir numerous hints for lines of experimentation that he might try.

All this time, Loir was working on his doctoral thesis. Entitled 'Microbiology in Australia', it would cover all his experiments in Australia, and would become the first study published in Europe about microbiology on the great southern continent. In his thesis, Loir would describe all his work in New South Wales and Queensland, covering chicken-cholera, anthrax, pleuro-pneumonia, rabies and glanders, in detail. Back at the Pasteur Institute in Paris, where he worked for several weeks with Emile Roux on research into tuberculosis in cattle, Loir ran his draft thesis by Pasteur.

It seems that Pasteur wanted to turn his nephew's thesis into a condemnation of the way that he, Pasteur, had been treated by the Rabbit Commission, the New South Wales Government, and others. The lengthy introduction to the chicken-cholera section rings with language that is typical of the by-now embittered and quite unwell Louis Pasteur, and very untypical of the polite, ultra-diplomatic Adrien Loir. It is clear that Pasteur dictated this part of the thesis to Loir.

In some parts, the thesis makes assertions that Loir would have known to be untrue; at the very least, they misconstrued the facts. For example, it blamed the squatters of New South Wales for banding together against Pasteur's chicken-cholera rabbit remedy so that they would continue to receive subsidies from the government. This, and similar beliefs about greedy colonial farmers, had permeated Pasteur's letters in 1888 and 1889. In an April 1888 letter, Pasteur had complained about people who would do anything and everything to obtain his chicken-cholera method without paying for it, describing Australians as 'a people greedy for money'.[381] On 29 June that same year, Pasteur had written to Mrs Priestley in England that 'my assistants will have seen the worst aspects of humanity and of men without honour'.[382] Two weeks later, in a letter to Loir, Pasteur was listing the enemies in Australia that his nephew must be wary of, including the members of the Rabbit Commission and 'even the squatters'.[383]

Loir had been on the ground in New South Wales, and had learned first-hand that the squatters had a passionate dislike for the enforced rabbit trapping program; after all, they had to pay for that program and then claim back a two-thirds subsidy from the government – it cost them thousands of pounds and did not solve the rabbit problem. This dislike of the subsidised rabbit progam had came out in testimony from many witnesses at Rabbit Commission sittings, and Loir was to talk with hundreds of farmers during the two and a half years that he was in New South Wales: from them he was able to gain the true picture. Fifteen thousand kilometres away on the other side of the world, Pasteur never could appreciate either the geographical or political situations in Australia, and it is his viewpoint, not Loir's, that comes across in Loir's thesis. Once again, Loir found himself in his uncle's power. Or, as he was to write some years later, in Pasteur's shadow. Feeling powerless to contradict Pasteur, he was compelled to let his uncle's observations on the chicken-cholera program stand.

Loir was to deliver the thesis in the second half of June. With this schedule in mind, he booked his return passage to Australia, departing from Marseilles in the first week of July, for he wanted to return in time to produce anthrax vaccine for Arthur Devlin to use in the new season's winter anthrax vaccination program. In the meantime, Loir had a luncheon appointment to keep. At the end of April, Sarah Bernhardt returned to Paris after a non-stop season in the United States that had begun the very same day she had arrived in San Francisco from Sydney, via New Zealand, aboard the *Mariposa*. Sarah was due to perform a Paris season from 1 May to 23 May and now, Adrien Loir would write, she made time for lunch with him.[384] At that lunch, it seems that the Divine Sarah told Loir that if he was determined to return to Australia then he must take a French bride back with him. It was not a wish, it was a command.

On 22 June, Loir presented his thesis at the Faculté de Médecine de Paris. After collecting his doctorate in medicine, he then took the train to Bordeaux, where he was soon joined by his parents. A week later, on 29 June, in a Roman Catholic ceremony at the 12th century Basilica of Saint Seuron at Bordeaux, Adrien Loir married Marguerite Morache.

Twenty-five-year-old Marguerite was the fourth and youngest child of Marie and Dr Georges-Auguste Morache. Dr Morache, Professor of Medicine at Bordeaux University and Director of the Health Service of the 18th Corps of the French Army, had led French missions to China and was a well-published authority on pathology, hygiene and military medicine. Morache had studied at Strasbourg University between 1845 and 1850, and there he had met his bride to be, Marie Caillot, daughter of a professor at the university. At Strasbourg too, Morache had befriended the family of university rector Professor Laurent. When Louis Pasteur had married one Laurent daughter and Adrien Loir Senior another, Morache had become friends with both the Pasteur and Loir families.

Yet even though the Morache and Loir families were old friends, Marguerite Morache knew little about the Loirs' son when he visited her parents' Bordeaux home in late February and early March of 1892. Intelligent, attractive but no beauty, Marguerite had long brown hair that she had not cut since she was a child – it hung down beyond her waist in a single plait. Her parents had raised her in 'a refined millieu', carefully planning her education so that she took advantage of her talent as a pianist and of her fine singing voice.[385]

With Adrien's May proposal of marriage coming like a bolt from the blue, the stunned Marguerite hesitated to accept at first.[386] But her suitor was the nephew of Louis Pasteur, he was about to qualify as a doctor, and he had great prospects in the world of microbiological research, so her parents encouraged the match. When Marguerite finally accepted Adrien's proposal, it meant that all the arrangements had to be done in a rush – the wedding invitations were issued on 15 June just two weeks prior to the wedding – for Loir still intended sailing for Australia at the beginning of July.

Following the nuptials, the couple hurried to Bordeaux railway station and caught a train from Marseilles. There, they would meet up with Dr Louis Momont, the nephew of Emile Roux who had originally been slated to go to Australia with Loir four years earlier. Momont had been working as a laboratory assistant with the Pasteur Institute since it opened, and Pasteur was sending him to Sydney to act as Loir's deputy. In Marseilles, Dr and Mrs Loir and Dr Momont boarded the Messagéries Maritimes steamer *Australien*, which sailed for Australia and Noumea on 3 July with a large complement of French passengers including a company of French soldiers bound for duty in New Caledonia. Marguerite Loir's honeymoon was to be a long ocean voyage, on the way to her new home in Sydney.

So it was that Adrien Loir returned to Australia with a bride, an assistant, his doctorate, a new moustache, and with news from New South Wales that the tide was at last turning in favour of Pasteur's rabbit remedy.

27

A CHANGE IN THE TIDE

ON 7 AUGUST 1892, a cool day in the late southern winter, the 6428-ton French mail steamer *Australien* docked at the Messagéries Maritimes line's wharf in Sydney, and the new Madame Loir surveyed the city that was to be her first marital home. Met by a coterie of Loir's Sydney friends as they came ashore, the Loirs and Louis Momont were conducted to the grand new Hotel Metropole, which had recently opened on the corner of Phillip and Bent streets. Opposite the Union Club, and just around the corner from Albany Chambers, the six-storey Metropole was in the heart of Loir's old stamping ground.

News of Loir's arrival was promptly reported in the press of several colonies, and he may have been surprised to find that the celebrity status he had achieved via the rabbit competition and the anthrax trial had not diminished. 'To all Australians,' wrote Brisbane's *Queenslander* in the third week of August, 'particularly those connected with pastoral pursuits, the name of M. Adrien Loir is now familiar as a household name. His many friends will be glad to hear of his safe return to Sydney after a visit of eight months to his native country.'[387]

Within several weeks of Adrien Loir's arrival back in Sydney, a reporter from *The Australasian Pastoralists' Review* tracked him down at the Hotel Metropole for an interview. 'Do you intend to settle here, permanently?' the reporter asked Loir.

'Yes, if I can obtain sufficient encouragement in the practice of my profession,' Loir replied. 'I have brought back with me, as partner, Dr L. Momont, who is also a member of Pasteur's Institute at Paris.'[388] As proof of his intention to stay, Loir had already registered with the New South Wales Medical Board as a legally qualified medical practitioner.

The subject of rabbits, the rabbit prize, and Pasteur's chicken-cholera entry came up very early in the lengthy interview with the *Review* reporter for, as Loir had heard while still in Paris, the mood in some New South Wales circles had turned very much in favour of Pasteur's rabbit remedy. In March, at the town of Nyngan in central-western New South Wales, a three-day conference of farmers large and small, who between them ran four million sheep and 10,000 cattle, had debated solutions to the continuing rabbit plague. The Nyngan conference had resolved to strongly urge the government to allow open country experiments with Pasteur's chicken-cholera microbes, and to renew the offer of a £25,000 reward.

Press reports of this conference stimulated ongoing discussion and correspondence. Frank Abigail, the former Parkes Government minister who had set the rabbit destruction competition in motion, wrote to the Sydney *Evening News* complaining that the Rabbit Commission had cost as much as £14,000 ($5.6 million) for no result, but agreed that the rabbit problem was 'one of the important national questions'.[389] Dr John Creed also weighed into the renewed discussion, reminding *Daily Telegraph* readers that he had been the first to conceive of the idea of a rabbit eradication contest, adding, 'I have never wavered in my opinion in this matter, and think that M. Pasteur's representatives were badly treated in not being allowed to use on a large scale his suggested chicken-cholera.'[390]

Several papers reported that at a September meeting of the Royal Society of New South Wales, Professor Thomas Anderson Stuart, head of Sydney University's medical school, denounced the idea of a rabbit eradication prize. 'Both Pasteur and Koch, in repeated conversations with me during the past winters,' Anderson Stuart told the Society, 'spontaneously expressed themselves in exactly the same terms with regard to this question of a prize.' Anderson Stuart said he was not against a biological remedy but he dismissed the Pasteur method. 'Although "fowl-cholera" has failed, still, that by no means is to say that other microbe diseases would also fail.'[391]

This sponsored the editor of the *Sydney Morning Herald* to haughtily disagree with Professor Anderson Stuart's prognosis, and to encourage Loir to persist with the Pasteur method of rabbit eradication: 'We are not disposed to accept Professor Anderson Stuart's dictum that fowl-cholera has failed. So, far from it, we would wish that the researches carried out by the learned pupil and nephew of Pasteur in our midst might be continued with vigour under government protection and sympathy.'[392]

It was in this atmosphere of considerable renewed support for a biological remedy, and for Pasteur's method in particular, including talk of the £25,000 prize, that the interviewer from *The Australasian Pastoralists' Review* quizzed Loir about what had gone wrong when he had first attempted to prove the efficacy of chicken-cholera in New South Wales. In response, Loir patiently recapped the whole sad story of his battles with the Rabbit Commission. When he mentioned that Pasteur had refused him permission to carry out the anthrax vaccination progam while he waited for the go-ahead with the final chicken-cholera experiment in open country, the reporter asked, 'What was his reason?'

Loir responded, 'Well, I suppose he was annoyed at the refusal of the government to allow us to deal with the rabbits in open country.'[393]

The conversation between Loir and the reporter moved on to Loir's work during his sojourn in France, and Loir told of how he had brought back to Australia tuberculine, a 'revealing agent' developed by Emile Roux that allowed swift diagnosis of tuberculosis in cattle; and Malleine, which allowed glanders to be identified within ten hours – not the ten days it had taken Loir the previous year with the Sell Brothers' circus pony. Loir also spoke of his plans to trial Professor Arloing's blackleg vaccine in New South Wales, and to follow up on the professor's suggested lines of pleuro-pneumonia research. Loir sounded excited, and looking forward to his new work in Australia.

Before the reporter ended the interview, he returned to the subject of rabbits and Pasteur's chicken-cholera scheme, throwing Loir a parting question: 'You have not heard whether the present government will allow you to conduct experiments on rabbits in the open country?'

Loir shook his head. 'No, up to the present time I have received no encouragement on that subject.'[394]

As Loir knew, the political landscape of New South Wales had recently changed. Sir Henry Parkes and the Free-traders were no longer in power, the last Parkes Government having fallen the previous October, and Loir had now to deal with a new government and a new slew of ministers. Ironically, it was federation that had been Sir Henry's undoing. Parkes had been driving the case for the federation of all the Australasian colonies, including New Zealand, into a single nation, and following his Tenterfield Oration of October 1889, he had called on his fellow premiers to send delegates to a Federal Convention that would put together a national constitution. The delegates to that convention, held in Melbourne in February 1890, had agreed to go back home and introduce legislation in every colonial parliament to permit federation to occur. A federated Australian nation seemed inevitable, as did Parkes' place at the head of it. This was all reported in Britain, from where Lord Tennyson had written to Parkes, 'We all hope to see you the first premier of the Australian Dominion.'[395]

By October 1891, the New South Wales federation bill was ready. But Parkes, the architect of Australian federation, then unaccountably backed away from his own design. He knew that a majority of his own cabinet had come to oppose the introduction of the federation act, preferring instead to move on issues such as 'one man, one vote'. Parkes had a firm policy of cabinet solidarity that prevented him from ever publicly criticising cabinet colleagues, so Free-traders in the Legislative Assembly – who supported federation by a large margin – could not fathom his apparent change of heart, and rebelled. A division in the House was decided only by the vote of the Speaker, and Parkes, peeved at his own members' lack of faith in him, resigned and went to the people. He lost the subsequent election to the Protectionists after many pro-federation voters turned against him because he had apparently shied away from federalism.

Parkes himself was reelected, but many of his followers including Daniel O'Connor, the colourful Irish Postmaster-General who had made the memorable welcome speech to Sarah Bernhardt, lost their seats. The Protectionists under Sir George Dibbs again took power, but this election had an unexpected side-effect: a large number of seats were won by the new Labor Party. Labor's success had much to do with legislation passed during Parkes' last ministry – legislation that he had personally opposed – to give New South Wales parliamentarians a salary. This had enabled many working-class candidates to give up their jobs to run on the Labor Party ticket. The day of the party machine had arrived, and Sir Henry Parkes would never again see the inside of the Colonial Secretary's office.

But by the time Loir arrived back in Australia, seventy-seven-year-old Sir Henry did have something to celebrate: on 2 August, five days before the *Australien* docked with Loir, his new wife and Momont aboard, the young new Lady Parkes gave birth to Sir Henry's latest child, a boy, at Hampton Villa in Balmain. 'There is no doubt that he is prouder of his latest achievement than even of his decoration, KCMG,' commented the *Illustrated Sydney News*, the latter referring to Sir Henry's second knighthood from the Queen four years earlier.[396] When Nellie and Sir Henry named their new son Cobden, a friend asked why they had not chosen a name such as Washington or Lincoln. Sir Henry had replied, 'Lady Parkes and I have thought of those names, and we are keeping them for the *next* three boys!'[397]

But all the joy of fatherhood could not dispel Parkes' gloom in the long term, for he was now a powerless backbencher, and he kicked in the traces of obscurity. The leadership of the Free-traders had passed to pudgy lawyer George Reid, and of the federation movement to Parkes protégé Edmund Barton; one would become the next Premier of New South Wales when the Dibbs Government fell in 1894, while the other would assume Parkes' predicted mantle, becoming the first Prime Minister of the new nation of Australia in 1901.

The Lands Minister in the Dibbs Government when Loir returned to Sydney in 1892 was Henry Copeland. Rabbits and Rodd Island became Copeland's ministerial responsibilities, and Loir needed the minister's permission to reoccupy Rodd Island. While Loir had been back in France, Tiger Bruce had written to him on behalf of the Mines Department, asking that the cost of sheep vaccination be reduced from three pence a head to two pence, and confirming that Loir could again use Rodd Island on his return to Sydney 'on the same terms as hitherto, for preparing vaccine for our sheep, or doing other work for this or any other department in this colony'.[398] To secure the island rent-free, Pasteur had agreed to reduce the price of anthrax vaccinations in Australia, but when Loir arrived back in Sydney in August he found that Rodd Island was the subject of a tug-of-war.

The Mines Department only managed Rodd Island on loan from the Lands Department, and while the Mines Department wished to honour Bruce's commitment to Pasteur and Loir, other parties were proposing different plans to Lands Minister Copeland for the island. Board of Health President Dr Frederick Manning – Dr MacLaurin's successor – had applied for the Rodd Island facility to replace William Hamlet's cramped Government Laboratory in the city. And five new Labor members of parliament had banded together to urge Copeland to return Rodd Island to public recreational use, a submission backed by a petition from residents of neighbouring Leichhardt.

For weeks and then months, Copeland put off a decision on the island's future, forcing the Loirs and Momont to continue living at the expensive Hotel Metropole. This would have suited young Marguerite Loir. The plush new hotel would have been very amenable to the daughter of a university professor, especially after sailing out from France in First Class luxury. The Loirs soon befriended other Metropole guests, including twenty-four-year-old Bertie Glasson and his young wife. Glasson, son of a wealthy squatter from Carcoar in the New South Wales central tablelands, lived permanently in a suite at the Metropole and threw money around as if it were going out of style.

The hotel's central location also meant that Marguerite Loir was within easy reach of shops, theatres and restaurants – especially the new Paris House Restaurant just a short stroll away in Phillip Street. Run by Gaston Lievain, a native of Lille, and his Belgian wife, the Paris House offered a ground-floor bistro and elegant upper-floor dining rooms, with a top-floor restaurant sponsored by Möet champagne whose menu included imported foie gras from Strasbourg, venison from Britain, and caviar from Russia.

While waiting for Minister Copeland to make a decision on Rodd Island, Loir, using Tiger Bruce's January letter as his authority, utilised the island's laboratory during daylight hours to manufacture anthrax vaccine for Arthur Devlin's latest vaccination program, which was already late in commencing. Loir and Momont daily took a launch out to the island and quickly cultured a batch of vaccine – their first advertisement for vaccine sales was running in the *Sydney Morning Herald* by 19 August, just twelve days after their arrival in Sydney. But living at the Hotel Metropole and travelling out to the island each day was proving an expensive and inconvenient way to relaunch the Pasteur Institute in Australia.

Loir also wanted to push forward with development of Professor Arloing's blackleg vaccine, and quickly arranged for a trial on William A. Walker's property at Tenterfield in northern New South Wales. Leaving Momont behind to continue vaccine production, Loir and his new wife took the train north, with Loir looking forward to showing Marguerite the interior of this country that had become his second home. On 15 September, the Loirs arrived in Brisbane and booked into the swish Gresham Hotel, having conducted the first stage of the blackleg trial at Tenterfield the previous day. Loir had written to the Premier of Queensland, Sir Samuel Griffith, on 12 September, and the day after he arrived in Brisbane he met with the Premier. The Brisbane *Evening Observer* commented of Loir, 'He regards the object of his visit as confidential at present.'[399]

Loir would have discussed three things with Premier Griffith. One topic would have been his continuing pleuro-pneumonia and blackleg projects; another would have been his 1889 recommendation that a stock institute be established in Brisbane. Thirdly, he pushed the subject of chicken-cholera as a rabbit eradicator. The *Evening Observer* had interviewed Loir at Tenterfield on 14 September, just before his departure for Brisbane, and had reported, 'He asserts that the inoculation of rabbits with chicken-cholera, which is a disease not liable to be contracted by any farm animal except poultry, would effectually banish this pest.'[400] Loir was as determined as ever to push for the adoption of Pasteur's rabbit remedy, and he found Premier Griffith to be of the same mind: shortly after this meeting, Griffith declared publicly that he 'favoured further experiments in M. Pasteur's method of rabbit inoculation'.[401]

In the latter part of September and early October, Loir successfully completed the blackleg trial at Tenterfield. He then hurried back to Sydney with Marguerite, for he had learned that Minister Copeland planned to take a party of parliamentarians and reporters on an inspection tour of Rodd Island. When, on Saturday 8 October, Copeland – a very large, rough and ready Yorkshireman – arrived at the Rodd Island jetty in a steam-launch with his party, Loir, Momont and Marguerite were all there to greet them. 'Dr Loir and his charming wife, who have only been a few months in the colony, entertained the company,' reported the Sydney *Daily Telegraph*, 'and with also Dr Momont's ready assistance, explained the work that was proceeding on the island.'[402]

Loir's public relations exercise worked – Copeland announced to the pressmen in the party that he did not propose to change the present arrangements regarding the island, and the following Tuesday he told the Legislative Assembly that it had been decided 'to allow the island to be used for scientific purposes for the present'. The government had made this decision about Rodd Island, he said, 'in order that Dr Loir might have the use of it for the conducting of his important investigations, and the preparation of Pasteur's vaccine of anthrax.' Placating the Labor members, he added that it was 'anticipated that there would only be a temporary delay' before the island became available for public recreation.[403] The next day, Loir, Marguerite and Momont checked out of the Hotel Metropole and moved onto Rodd Island and into the residency.

When Minister Copeland made his 11 October announcement about Rodd Island, he also said that the arrangement allowing Loir to use the island was 'pending the voting by parliament of money for the establishment of an Intercolonial Stock Institute'.[404] Next day in the House, Protectionist member Dr Andrew Ross picked up on this remark and was told by Mines Minister Thomas Slattery that the government planned to set up an Intercolonial Stock Institute, a research body based on the recommendations of the November 1889 Australasian Stock Conference in Melbourne and later promoted by Sir Henry Parkes. Initially, said Slattery, this new institute would probably be located on Rodd Island, but the plan was to relocate it to a site associated with Sydney University.[405]

With the New South Wales press applauding the Australasian Stock Institute scheme, and Sydney University enthusiastically getting behind it and developing plans for its operation as both a research and teaching institution, the Dibbs Government approached the other colonies, proposing that they all contribute to the institute's funding based on the number of stock in each colony. At the same time, it was an open secret that it was intended to offer Dr Adrien Loir the post of Director of the institute. Whether Loir encouraged this, and whether he was genuinely interested in the post, is unclear. In the end, only three of the seven colonial Governments – New South Wales, Queensland and Tasmania – agreed to support the institute; Victoria was apparently not interested because it would be located in New South Wales. Despite the lack of support from the other colonies, the three supportive governments decided to move forward with the project.

For Loir, everything seemed to be going right. The Stock Institute project was progressing, Loir had secured Rodd Island and had resumed residence and experiments there, Arthur Devlin had the season's anthrax vaccination program in full swing, and the mood in New South Wales and Queensland was increasingly in favour of giving Pasteur's chicken-cholera rabbit remedy another try. And, to Loir's joy, Marguerite announced that she was pregnant; the child she was carrying had been conceived on or close to their wedding night.

But then, something occurred that changed everything for Loir in Australia. And it was all to do with Louis Pasteur's bitterness at his failure to date to be awarded the rabbit eradication contest prize.

Loir's doctoral thesis, *Microbiology in Australia*, had been published in Paris as a slender little book. By September, copies had started to arrive in Australia and the local press set about having it translated. The colonial reaction to Loir's thesis was mixed, ranging between surprise, alarm and outright anger. *The Illustrated Sydney News*, which the previous November had published a major feature about the work of the Pasteur Institute in Australia, was among the newspapers that fell into the alarmed category: 'He plainly states that he was unable to properly test the Pasteur rabbit extermination scheme by reason of the opposition of the squatters themselves,' the paper now said of Loir.[406] This claim was Pasteur's contribution to his nephew's thesis, and it was coming back to bite him and Loir. 'He has little good to say of the "personnel" of the Rabbit Commission,' the *News* went on, 'and declares that the only man who held out a helping hand to him or his scheme was the minister. Now these accusations are grave, and the government should look into the matter without delay.'[407]

At the angry end of the scale, the incensed Sydney *Daily Telegraph* reported that Loir had made a 'rather remarkable accusation against the runholders of New South Wales'. The *Telegraph* huffed, 'The scheme of rabbit destruction by chicken-cholera, it is gravely stated, was rejected because it was recognised that if once tried it would inevitably prove successful. This hypothesis is no doubt a very flattering unction for Dr Loir to lay over his own soul, but will hardly find very general endorsement.'[408]

Other publications such as *The Australasian Pastoralists' Review* and the Brisbane *Evening Observer* published large slabs from Loir's text, and left it up to readers to make up their own minds about what they thought of it, and him. Brisbane's *Queenslander*, which had been so pleased to hear of Loir's return to Australia in August, was a little shell-shocked by the controversial claims contained in his thesis, and told its readers that 'Dr Loir expresses the opinion that the great opposition manifested towards M. Pasteur's scheme was solely owing to the station-holders in the infested districts.'[409]

The *Queenslander* also mentioned that Loir had 'much fault to find' with Rabbit Commission personnel, and that he had also blamed the influence of Victorian interests in the outcome of the Commission's findings.[410] There can be no doubt that these two comments can be genuinely attributed to Loir rather than to his uncle, as both claims were very accurate. The reference to the Victorian influence on the outcome of the Commission was particularly true – the Commission's President, Professor Harry Allen, who had been against Pasteur's proposal from the first day, was a Victorian, while the two other Victorian commissioners, Lascelles and Pearson, had steered the Commission toward recommending that rabbit-proof fences be made compulsory as opposed to employing a biological remedy.

However, another claim contained in Loir's thesis and reported by the *Queenslander* had the Pasteur stamp to it. This was a claim that Frank Hinds had not been well received in Australia because 'the selection by M. Pasteur of an Englishman was looked upon as an outrage to Australian independence and treated as such'.[411] This reflects comments made by Pasteur and his English friend Eliza Priestley in letters they exchanged in 1888 and 1889 and in Pasteur's letters to Loir. In an April 1888 letter, Pasteur had described Australians as 'jealous of anything foreign'.[412] In his 1889 interview with *Le Siècle*, Pasteur had accused the Rabbit Commission of failing to accept his chicken-cholera method 'so as not to award the prize to a foreigner', and had accused the Australian medical community of fearing 'to look to Paris and be denounced by some of their compatriots in English medicine'.[413] It was Pasteur not Loir who had formed the opinion that he'd been deprived of the prize because of Australian xenophobia. Apart from in his thesis, nowhere did Loir write or was quoted as saying anything remotely like this, either during his years in Australia or later.

Loir was to say that Hinds told him his English medical qualifications had stood in the way of his obtaining a position with an Australian hospital,[414] but the claim in his thesis that Hinds' inclusion in the Pasteur Mission was 'an outrage to Australian independence' was almost certainly inserted by Pasteur. Not only had Pasteur repeatedly accused Australians of xenophobia, he would have felt compelled to defend his faulty choice of Hinds as the mission's interpreter, and the most convenient way for Pasteur to excuse Hinds' failure in this role was to blame the Australians, not himself.

If Loir's thesis had simply confined itself to lambasting the compromised Rabbit Commission and its flawed report, Loir might now have received a more favourable reaction in New South Wales. But Louis Pasteur's determination to put his imprint on his nephew's paper, using it to defend himself against critics in France who could not understand why he hadn't been awarded the rabbit prize when he had seemed so certain of it, now backfired on both uncle and nephew.

Not only were there adverse comments about Loir and Pasteur in the colonial press, in November members of the New South Wales Legislative Assembly called for a royal commission into the proposed Australasian Stock Institute before it was allowed to go any further, implying a searching inquiry into Dr Loir and his character. The institute project came to a shuddering halt. And overnight, all the talk of the previous months about supporting an open country trial of the Pasteur chicken-cholera remedy, and the recommended payment of a £25,000 reward, dried up, both in the press and in the parliament.

If Loir felt this was a bad period for him, his one-time supporter Frank Abigail was having an even more torrid time. The former Mines minister had become chairman of the Australian Banking Company and, like many a bank and building society in the 1890s, it collapsed in 1892. As a result, in October Abigail and six other directors were brought to trial on criminal charges of falsely representing the bank's affairs. After being found not guilty, Abigail was charged with conspiring to issue a false balance sheet. Found guilty of this charge on 3 November, Abigail was sentenced to five years' hard labour in Darlinghurst Gaol – breaking rocks in a chain gang – and was taken away in chains.

In December, with the Australasian Stock Institute stalled by calls in New South Wales for a royal commission, the Queensland Government announced that it was going to create a stock institute of its own, and that it would offer Dr Adrien Loir the job as its Director. Nothing more was heard as Loir, his wife and Momont spent Christmas in Sydney. But then in January, Loir unexpectedly announced that he and his now five-month-pregnant wife were returning to France for family reasons. Loir appointed Momont Acting Director of the Pasteur Institute of Australia and then, giving every indication that he intended to soon return to Sydney, took a First Class stateroom on a French steamer bound for Marseilles. Dr and Madame Loir then quietly sailed away. By the end of February they were back in Paris.

In February, the Queensland Government confirmed that it was going ahead with its stock institute. On 18 March, the *Queenslander* reported on a meeting of the Queensland Executive Council which had discussed the appointment of a Director for the planned institute: 'It has been agreed to offer the position to M Loir, who is now on a visit to Paris.' The paper had only one concern: 'Will the salary offered to M. Loir be sufficient to induce a scientist so distinguished to become Director of the institute?' The salary of £500 a year with a two-year contract was contained in a letter of offer that was immediately sent to Loir at the Pasteur Institute in Paris.

In Sydney, *The Australasian Pastoralists' Review* praised the Queenslanders for their initiative, and bewailed New South Wales' apparent loss of Dr Adrien Loir to Queensland, complaining that such an institute had been on track for Sydney. 'Some unintelligent members of the Legislative Assembly, however, poked their interfering noses into the matter, and tabled such insulting and ridiculous questions in the House on the matter that Dr Loir, disgusted beyond measure at the lack of appreciation his gratuitous efforts aroused, packed his trunks and returned to Paris a couple of months ago.'[415]

More months then went by without a word from Adrien Loir. It is clear now that Loir was going through one of the most difficult periods of his life. His young wife Marguerite had lasted barely six months in Australia. After the initial excitement, the travel, the expensive hotels and restaurants, she had found herself on a little island with no one to talk to other than her husband, his assistant, and the domestic staff. Adrien would spend long hours in the laboratory, totally obsessed with his work, just like his uncle. He was of course back in his element, with his own island, his laboratory, and pottering around Iron Cove in his boats, fishing and duck-hunting.

But this was not a lifestyle to which Marguerite was accustomed. Marguerite did not speak English and had never even been away from home before, while her husband had been to Russia to help the Czar, had been to Copenhagen to make Danish beer, and had spent five years in Australia becoming acclimatised to the country, to the foreign flora and fauna, and to the Australian people. Marguerite did not even have a piano on Rodd Island with which to while away the hours. Rodd Island, Loir's paradise, became Marguerite's prison. And there was another negative element: according to Loir's granddaughter Françoise Michel-Loir, on Marguerite's arrival in Sydney she discovered that Adrien had many female friends in New South Wales, friends who still wanted to share his company; this did not go down well with Loir's insecure new French bride.[416]

Soon jealous, lonely, and desperately homesick for her family, her home town, and the French way of life, Marguerite must have been frequently in tears. Yet a frustrated Loir could not offer a solution. This was where he wanted to be. The Christmas of 1892 in Sydney must have been the final straw for Marguerite – the first Christmas she had ever spent separated from her mother and father. Totally miserable, uncomfortably pregnant in the Sydney heat, and unable to take any more of this unfamiliar and in some ways threatening land, it seems that she demanded that Loir take her home at once, so that she could give birth to their child in France.

Back in Paris in the northern spring, Loir was pining for Australia and for the opportunities he had left dangling there. He sat down and wrote an article about the Aborigines of Australia for the leading French scientific journal *La Nature*, going into considerable detail about tribal customs and providing illustrations based on his own photographs; the article was published in two parts in the journal's June and July issues. He also wrote an article for the 20 April issue of *Revue Scientifique* entitled 'Rabbits in Australia', in which he recapped the whole chicken-cholera saga and said that he still believed the Pasteur proposal was the best remedy for Australia's rabbit plague. It is evident that, in his heart, Loir felt he could prove Pasteur's method to the Australians if given the chance.

In May, in Bordeaux, Marguerite Loir gave birth to a son whom she and Adrien named André. But when Loir now spoke of returning to Australia, he found that not only was Marguerite opposed to it, even his colleagues spoke against the idea; indeed, Emile Duclaux, now Louis Pasteur's right-hand man, counselled Loir not to return. Apparently Loir had not revealed to his Pasteur Institute colleagues that he had received a job offer from the Queensland Government, for Duclaux told him that the head of the Pasteur Institute of Australia should have competencies in medicine, veterinary science and chemistry that Loir did not possess. He advised Loir to gain more experience, and to forget about Australia. Today, Loir's granddaughter Françoise Michel-Loir poses the question: was Duclaux offering impartial advice or was he jealous of the success and notoriety that Pasteur's nephew had achieved in Australia?[417]

That May, Loir finally sat down to reply to the Queensland job offer. Even if he had been stung by the criticism he had received in New South Wales after publication of his thesis, he was still well-liked in Queensland where he had many friends and supporters, not the least in government. Loir loved Australia, of that there was no doubt, and he loved the freedom he had known there. To outsiders, there would have seemed every reason to accept the Queensland offer, and little reason to turn it down. But there were several governing factors: Marguerite would not return to Australia under any circumstance, Duclaux was dead against it, and accepting the Queensland offer meant leaving the Pasteur fold. In the end, Loir could not desert Louis Pasteur. Or was not permitted to.

The Queensland job offer had been conveyed to Loir by F.A. Blackman, Honorary Secretary of the Queensland Stockbreeders and Graziers Association, so Loir addressed his letter to him. 'Unfortunately,' he wrote, 'family reasons oblige me to decline at the moment the offer of a position for which I worked so long in Australia.' Through Loir's palpable regret shone a dim ray of hope that perhaps the situation might change before long, and somehow he would be able to accept the post after all. Like a man parting from a dear one whom he feared he would never see again, Loir clung to the outstretched Australian hand, adding, 'If I myself am not to be the head of this institution, I want to remain a constant correspondent.'[418] Later events suggest that Loir did indeed correspond with the man who ultimately took up the post of Director of the Queensland Stock Institute.

To take the gloomy Loir's mind off Australia, Emile Duclaux suggested he accompany him on a trip to Tunisia in North Africa. Duclaux was heading off to study grape fermentation on behalf of the wine industry there. So, leaving Marguerite and baby André in Bordeaux, Loir sailed for Tunis with Duclaux, and tried to forget about Queensland and his dashed Antipodean hopes and dreams.

The pair's work in Tunisia was soon so productive that the locals asked for a branch of the Pasteur Institute to be set up in their country, and when Duclaux returned to Paris he left Loir behind to establish and run a new Pasteur laboratory in Tunis. Loir now sent for Marguerite – at least people in Tunisia spoke French, so Marguerite would feel more at home. Loir's wife and baby son duly joined him in Tunis. In the Villa Pasteur, a grand house by the harbour in the city's marine quarter, Loir set up a laboratory and Marguerite set up a piano; before long she would make the Villa Pasteur the centre of the French community's social life in the city.

With his champion withdrawing from the rabbit race, Louis Pasteur's bid for the Australian rabbit eradication prize finally and abruptly came to an end. There can be little doubt that had Loir's thesis not been published in Australia, and had the thesis not contained criticisms including a claim that Australian farmers had conspired to prevent Pasteur from winning the competition prize – a claim that put both the press and parliamentarians of the colony offside – the renewed efforts that were at that time afoot in New South Wales to press for the introduction of Pasteur's method and award him the £25,000 prize would quite probably have been successful.

Despite Pasteur's failure to win the New South Wales rabbit destruction competition, his bid for the prize had borne fruit in other ways, for it allowed Adrien Loir to generate millions of francs for Pasteur in Australia from his other microbiological work. Exceeding the value of the competition prize, this Australian money had permitted the under-funded new Pasteur Institute to pay its way, thereby ensuring that it remained under Pasteur's control. But the story does not end here. Despite Loir's departure, the Pasteur Institute of Australia continued to operate, and before the story of Louis Pasteur's Australian gambit came to an end there would be several surprising twists in the rabbit tale.

28

THE FINAL TWISTS

From Tunis, where his Pasteur Institute of Tunisia laboratory consisted of two small, cramped rooms, Adrien Loir kept track of what was happening in Australia via letters from Louis Momont, Tiger Bruce, Arthur Devlin and his many other friends in New South Wales and Queensland. They would often send him newspaper cuttings about goings-on in the colonies, particularly rural matters, and he would diligently paste them into his latest scrapbook.

One of the news stories from New South Wales that went into Loir's scrapbook was about Bertie Glasson, the big-spending squatter's son whom the Loirs had befriended at the Hotel Metropole in Sydney. By September 1893, Bertie had run out of money and had gone back to his home town of Carcoar to rob the local bank. Caught in the act one night as he tried to batter his way into the strong-room with an axe, Bertie killed two people with the axe as he tried to get away; he was later hanged for the crime. Loir also heard that, with the Australasian colonies now plunged into an economic recession, his friend Walter Lamb, the Union Club member who had taken him to meet Frank Abigail in 1888 in the wake of the Marquis De Rostain's atrocious attempt at translation, had gone bankrupt in November when his ambitious Sydney cannery had dragged him under.

Loir also heard that, on 2 December 1893, Queensland opened its stock institute. In lieu of Loir, the Queensland Government had appointed twenty-seven-year-old Dr Charles J. Pound as the institute's first director. An Englishman who had studied at King's College, London, Pound had been enticed to Australia by Professor Anderson Stuart of Sydney University with a promise of the directorship of the Australasian Stock Institute in Sydney. Anderson Stuart had been convinced the Sydney institute would proceed after Adrien Loir left town, thinking Loir's parliamentary critics would then drop their opposition to it. He was proven wrong. C.J. Pound, finding that the Sydney institute still only existed in the minds of its proponents, was forced to work temporarily as a laboratory assistant for William Hamlet before he applied for the Brisbane position. Because he had received microbiological training at the Pasteur Institute in Paris – a requirement for the post laid down by the Queensland Government – C.J. Pound won the appointment.

Meanwhile on Rodd Island, Loir's former colleague Louis Momont, whom Pasteur officially appointed Director of the Pasteur Institute of Australia, manufactured anthrax vaccine and the Arloing blackleg vaccine. Momont was also able to offer Australian graziers a pleuro-pneumonia vaccine because Adrien Loir had finally made the breakthrough with the virus towards which he'd been working for years. The breakthrough had come while Loir was working with Pasteur and Roux at the Pasteur Institute in Paris in the spring of 1893, before he went to Tunis. This success earned Loir the University of Paris's Monbine Prize in 1893, and nomination for chevalier of the *Legion d'honneur.*

When Loir failed to return to Sydney, the busy Louis Momont needed assistance so he turned to a Sydney man, John McGarvie-Smith. Tall, barrel-chested, bearded, and immensely strong, the forty-nine-year-old McGarvie-Smith had been born plain John Smith before he took the name of his local minister, the Reverend McGarvie, to give himself a colourful moniker that matched his vibrant personality. A crack shot who in 1876 had led the first Australian shooting team to compete in the United States, McGarvie-Smith had worked as a metallurgist for some years before moving into microbiology.

In 1891, McGarvie-Smith had paid Dr Oscar Katz – the same Oscar Katz of Rabbit Commission fame – to train him in laboratory techniques for six months. Katz was by that time desperate for paid employment. He'd taken a hammering in New South Wales from supporters of vaccination, and in Victoria his reputation had been further dented when he'd failed to identify the typhoid bacillus in a sample of Melbourne water sent to him for testing. It was subsequently revealed that the sample had contained the typhoid bacillus, for it had been deliberately tainted with the bacillus by chemist Alphonse de Bavay, who had set out to embarrass Katz's employer, Professor Harry Brookes Allen, the president of Melbourne's Sanitation Commission.[419] Come the end of 1891, it seems that Katz had returned to Europe.

There was nothing that John McGarvie-Smith would not venture, nowhere he would not go: between November 1892 and July 1893 he conducted bacteriological surveys of the air in the sewers of Sydney for the Metropolitan Water Board on a £100 contract, descending by rope into the foul city sewers twenty times to take his samples. Since 1892, McGarvie-Smith had also tried unsuccessfully to create a snake anti-venine, personally catching five hundred poisonous Australian snakes and milking them of their venom. When Smith approached Louis Momont for advice on the snake project, Momont was impressed by the big Australian's enthusiasm and confidence, and in the second half of 1893 offered him the job of laboratory assistant. Before long, McGarvie-Smith would play a crucial role in the future of the Pasteur Institute of Australia.

In the southern summer of 1893–94, Louis Momont went home to Paris where he married Alice Debenay on 24 February 1894. He returned to Sydney, and Rodd Island with his new wife, and a new scientific partner who'd been trained at the Pasteur Institute in Paris, Dr Emile Rougier.

On 13 June 1895, Louis Pasteur left the Pasteur Institute at Grenelle for the last time. A carriage conveyed him and wife Marie into the Rue d'Utot, then headed west for Villeneuve l'Etang, just outside

Paris. Now that the Pasteur Institute had been operating for close to five years, Pasteur's money worries had eased. Millions of francs were coming in every year from the institute's overseas operations – from Louis Momont's Australian branch, from the branch run by Adrien Loir in Tunisia, and from a new Pasteur Institute of Viet Nam in French Indo-China. All were contributing significantly to the institute's coffers, as were vaccine sales in France, Spain, Italy, Austria, India and South America. Since the opening of the institute in 1888, generous private donors had also come forward, and in the next few years major benefactors would contribute to significant expansion at the Grenelle site. Despite his financial fears, once Pasteur had built his institute and showed what was possible, the money to support its research and teaching operations had been forthcoming, not the least from Australia.

Pasteur would have been feeling pleased with himself as the carriage rolled out into the French countryside that June day, and just a little smug; the previous month, the Berlin Academy of Sciences had announced its intention to award Pasteur the Prussian Order of Pour le Mérite for his services to science. Pasteur hadn't changed his attitude toward the Germans. With Alsace and Lorraine still in German hands, and with the Pour le Mérite presented by Kaiser Wilhelm II himself, Pasteur was not interested in the award. He had refused the German honour, and had enjoyed doing it. But Louis Pasteur now felt utterly worn out. Seeing this, Marie decided not to inflict on him the exacting journey to Arbois this summer. A few quiet months in the countryside just outside Paris, a removal from the bustle of the institute yet without the rigors of the trip to Arbois – this might be the thing to rejuvenate him.

The coach's destination at Villeneuve l'Etang was a vast park. There was a former imperial château in the park, once a favourite retreat of Napoleon III. Here too there was a stables complex that had formerly been occupied by the cavalry of the Imperial Guard. Since 1884, the Government had allowed Pasteur to use the stables to house animals involved in research projects; more recently, the Pasteur Institute had been keeping horses there, using them in the production of an anti-diphtheria serum. In a house on the verdant Villeneuve l'Etang estate, Pasteur was to pass the summer.

While Louis Pasteur settled into the peaceful countryside on the fringe of Paris, in Sydney Sir Henry Parkes was gearing up to fight his last election. The Dibbs Protectionist Government was at loggerheads with the Upper House, which refused to pass a bill that would impose land tax and income tax in New South Wales for the very first time; an election was the only way to resolve the impasse. Parkes had become so embittered by his usurpation as Free-trade party leader by George Reid that when the election was called, Parkes declined to renominate for his St Leonards seat and instead contested Reid's Sydney seat – even though Reid was now Premier, widely popular, and a fellow Free-trader.

Parkes' loyal young wife Nellie encouraged him to take on Reid even though she was by now desperately ill, suffering from a terminal condition – apparently cancer. Nellie's situation was not helped by the fact that the Parkes family had been evicted from Hampton Villa after falling behind in the rent. Parkes' income had been badly dented by the New South Wales recession, and he struggled to support himself, his wife, their children, her children, his children including two unmarried daughters and a third who was now widowed, as well as a governess and domestic staff, on just £212 a year. This was because the recession had more than halved the £540 in interest payments that had originally been generated by the Parkes benevolent fund. The Parkes tribe had moved into a cheaper rented house, Kenilworth, a Tudor-Gothic pile at Annandale in Sydney's inner west, and friends in government arranged for nurses to attend the dying Nellie around the clock.

On the stump, Parkes railed against Reid as though he were a Protectionist enemy. At home, Nellie, lying in her deathbed, maintained support for her husband to her final breath. 'Never surrender,' were her last words to him.[420] She died just prior to the end of the election campaign. This was probably a good thing, for it meant that Parkes did not have to face her with the news that he had been defeated by Reid, and that he was now no longer even a member of parliament – for the first time in forty years.

By September, as the northern summer came to an end at Villeneuve l'Etang, Louis Pasteur's health was declining rapidly. Unable to even walk in the park, he was confined to bed and Marie decided not to move him. Family members and his Pasteur Institute disciples came to sit around his bed. Jean-Baptiste read his father accounts of Napoleon's battles. Then Pasteur suffered another stroke. After that, the end was only a matter of time.

At 4.20 on the afternoon of 28 September 1895, clutching Marie by the hand, a crucifix in the other hand, seventy-two-year-old Louis Pasteur passed away, surrounded by those who loved him. His funeral, at Notre Dame Cathedral in Paris, was attended by the President of France, every cabinet minister, a host of foreign dignitaries, and the entire staff of the Pasteur Institute led by its new Director, Emile Duclaux, while the people of Paris reverently lined the streets to watch the great man's cortege pass by. He was laid to rest in a crypt in the basement of the Pasteur Institute at Grenelle.

Six weeks later, and four months after the death of his second wife, eighty-year-old Sir Henry Parkes astonished everyone who knew him by marrying for the third time. But no one was much surprised when they learned that his new bride, Julia Lynch, was a raven-haired beauty of twenty-three; Sir Henry might not have money or a seat in parliament but he still had charisma and optimism. In December, knowing that Sir Henry was in dire financial circumstances – there being no pension at that time for former members of the New South Wales parliament – the Legislative Assembly debated giving him a parliamentary allowance, but the House rose without taking a vote. By the beginning of April, bailiffs were banging on the door at Kenilworth to remove the furniture so it could be sold to pay Parkes' debts. Besieged by his creditors, Sir Henry fell ill and took to his bed.

In the last week of April, Premier George Reid called at Kenilworth, having only learned of the seriousness of Parkes' illness in the morning's paper. 'How is he?' he asked young Julia, the new Lady Parkes. 'I scarcely like to ask him to see me, if he is very ill,' he added.

Julia asked the Premier to wait a moment and went upstairs to Sir Henry's bedroom. But when Parkes learned that it was Reid who was asking to see him, he refused. It took Julia and Sir Henry's doctor some minutes to convince the old man to change his mind. Finally, Premier Reid was permitted to come in. When he saw him, Parkes weakly extended his hand. Reid crossed to the bedside, took the grand old man's hand, then bent and kissed it. Parkes bit his lip and tightened his grip. Tears began to form in George Reid's eyes. And tears rolled down Henry Parkes' cheeks. They did not say a word, just remained holding hands for a long, long time until Reid could bear no more and withdrew.

'I am glad he called,' Parkes said, as Julia dabbed his damp cheeks. 'I am glad I saw him. I have misunderstood him.'[421]

Two days later, at 4.00 in the morning of 27 April 1896, Sir Henry Parkes died. He had said that he did not want a state funeral, so his family arranged a private affair. But the people of Sydney would not let his passing go without a final salute. Australians value their sports people more than their politicians. When, in 1889, world champion New South Wales oarsman Jack Searle died in his prime at the age of thirty, his life claimed by that year's typhoid epidemic, 100,000 fans had crowded to his public Sydney funeral. Just the same, more than a few people also wanted to say goodbye to the grand old man of Australian politics: excluded from his funeral, 30,000 people lined the streets of Sydney to silently watch Sir Henry Parkes' coffin go by on its way to its final resting place at Faulconbridge in the Blue Mountains.

In late 1892, Parkes had published a two-volume autobiography entitled *Fifty Years in the Making of Australian History*. Many people had expected a lively read but, unlike the man himself, his biography proved to be deadly dull. A catalogue of political events, of countless campaigns and ever-changing ministries, it gave no insight into the man nor his colourful personal life. Sir Henry mentioned many interesting people he had known and dealt with, including Tennyson and Gladstone, but of Louis Pasteur he said not a word. The rabbit eradication contest did not rate a single paragraph, perhaps because Parkes had considered it a failure. He did not even take credit for attracting Pasteur's nephew to Australia, for facilitating anthrax vaccination in Australia via his covert action, nor for the resultant foundation of the Pasteur Institute of Australia and the vaccination benefits enjoyed by Australian farmers as a result.

By Parkes' death in 1896, anthrax vaccination was being employed by increasing numbers of stockowners in New South Wales who were enjoying the benefits of vigorous competition between two different vaccine suppliers – for the Pasteur Institute no longer had the market to itself. Alexander Gunn, the manager of Yalgogrin Station, a property owned by the Goldsborough Mort Company, had been one of the two hundred men in the crowd at Junee in October 1888 when Loir and Germont had demonstrated their anthrax vaccine. Impressed by the two Frenchmen and their vaccine, Gunn had been using Arthur Devlin's vaccinators since 1890. The Pasteur vaccine had a limited life and could only be used in colder months; even then, the glass tubes were rushed by Devlin's men from railway station to sheep run in fast buggies and wrapped in wet cloth to keep them cool. Despite these precautions, the Pasteur vaccine was not entirely reliable, and up to five percent of sheep inoculated with it died.

Alexander Gunn was sure that it was possible to produce a better vaccine, and he set out to prove it. Although Gunn had no medical or scientific training, he created his own laboratory on Yalgogrin Station, imported a Zeiss microscope from Germany, and began experiments. He was to claim his experiments were based on notes written by Adrien Loir and which had been passed on to him by Arthur Devlin.[422] With his wife taking precise notes of everything he tried, Gunn experimented for four years, aiming to create a vaccine that required just a single dose, as opposed to the two-dose Pasteur vaccine, and that would keep indefinitely.

In 1894, Gunn succeeded in creating and testing his one-dose anthrax vaccine. After much jubilant celebration in the Gunn household, he tried to repeat the process. But no matter what he did, Gunn could not reproduce the original vaccine. Unable to find the answer in his own notes, the frustrated amateur microbiologist approached Arthur Devlin for advice. Devlin, realising that he would be cutting his own commercial throat by helping Gunn come up with a vaccine that improved on the Pasteur product he was marketing, declined to help him. So in early 1895, Gunn discreetly approached John McGarvie-Smith, who was at that time a laboratory assistant at the Pasteur Institute on Rodd Island, for his help. The ambitious McGarvie-Smith agreed to see if he could reproduce Gunn's one-dose vaccine using Gunn's notes and his own knowledge of the Pasteur process.

While McGarvie-Smith continued to work for Louis Momont and Emile Rougier on Rodd Island, he surreptitiously applied Louis Pasteur's secret to experiments with Gunn's process. For there really was a secret to producing the Pasteur anthrax vaccine, and that was why no one else in the world had come up with a successful vaccine to compete with the Pasteur product. At the Pasteur Institute in Paris, Emile Duclaux had modified the anthrax vaccine production process over the years, and the Pasteur secret entailed attenuating the vaccine through a series of animals; in Paris, the last animal in the chain through which the anthrax bacillus was passed before it was safe for vaccination was an ape.

Within months, using his Pasteur Institute inside knowledge and Pasteur's secret, the clever McGarvie-Smith had rediscovered Gunn's formula and secretly produced a batch of the new, long-lasting one-dose vaccine. If such a thing were to occur today – akin to a Coca-Cola employee stealing Coke's secret formula – McGarvie-Smith would be sued for millions by the Pasteur Institute for applying their know-how to his own venture in competition with the Pasteur product.

Alexander Gunn excitedly tested the new vaccine at Yalgogrin in front of interested local sheep-owners, and it worked. In May 1895, Gunn began vaccinating his neighbours' sheep, using a two-shot version of the vaccine that McGarvie-Smith was now producing in his Woollahra home – Gunn was worried that farmers accustomed to the two-shot Pasteur vaccine would not trust an unknown's single-shot variant. Apart from the reliability of the new vaccine, which lasted for months in the tube without the need for cold storage, Gunn made it attractive to potential clients by undercutting the Pasteur Institute's price – he offered to vaccinate mobs of 15,000 or more for one and a half pence per head, as opposed to the two pence being charged by the French.

Between May 1895 and April 1896, Gunn's team of nine vaccinators inoculated close to half a million sheep. During the same period, the Pasteur Institute of Australia vaccinated only around 100,000 sheep. Gunn grossed £3183 ($1,273,200).[423] But he received a rude shock from John McGarvie-Smith. The man making the vaccine was not content to merely be paid as a sub-contractor. McGarvie-Smith, who possessed the Gunn-Pasteur formula, threatened to go into business for himself in opposition to Gunn unless he made him an equal partner. What was more, McGarvie-Smith wanted top billing. Reluctantly, Gunn agreed. The McGarvie-Smith and Gunn Company was formed, and went into very profitable business.

In 1896, once his company was established, McGarvie-Smith could no longer remain at the Pasteur Institute of Australia – which had just relocated from Rodd Island to the Hermitage, a property at Double Bay, after the Lands Department finally surrendered to pressure to give the island over to public recreational use. Adrien Loir's laboratory on the island was torn down, as were the stables and the 'birdcage'. The kitchen building was retained, and so was the residency. Loir's home for a number of years was subsequently gutted and turned into a dancehall which proved popular with US servicemen during World War II. Today, controlled by the National Parks and Wildlife Service, Rodd Island can only be visited with a special permit.

At the same time that McGarvie-Smith left the Pasteur Institute of Australia in 1896, Louis Momont went home to France, leaving Emile Rougier in charge of the Sydney institute. That 1896 season, McGarvie-Smith and Gunn introduced their single-shot vaccine which, by 1898, increased their annual income to £6500 ($2,600,000). That same year, the Pasteur Institute of Australia threw in the towel. Unable to compete with McGarvie-Smith and Gunn, it sold the Hermitage and its Australian vaccination business to Monsieur F. Zimmermann, who had trained in Paris at the Pasteur Institute. Zimmermann set up a company in Sydney, Pasteur Anthrax Laboratory Limited, to produce the Pasteur vaccine under licence, and was soon conducting a lively battle in the press with his competitors about the merits of the opposing vaccines. McGarvie-Smith and Gunn won the battle – Zimmermann ceased trading within two years while the two Australians became very wealthy men.

By the time of Alexander Gunn's death in 1910, he and McGarvie-Smith had not been on speaking terms for years. In 1917, McGarvie-Smith gave their closely guarded anthrax vaccine formula to the state of New South Wales, along with £10,000 ($4 million), on condition that it be used to establish a McGarvie-Smith Institute. John McGarvie-Smith was worth £28,739 ($11,495,000) when he died in 1918.[424] The McGarvie-Smith Institute still exists today, in Sydney, as a funding body for animal disease research conducted by Australian scientists. There are thirty Pasteur Institutes around the world, but thanks to John McGarvie-Smith there is no Pasteur Institute in Australia today.

There is one more twist in the rabbits' tale but, before we get to it, I should tell you about the fates of the two men chiefly responsible for the failure of Louis Pasteur to be awarded the rabbit extermination contest prize. Professor Harry Brookes Allen, the President of the Rabbit Commission who, along with Dr Camac Wilkinson, must share the major burden of responsibility for deliberately misrepresenting and derailing Louis Pasteur's contest entry, was knighted in 1914 on the recommendation of the Victorian Government. He continued as dean of the University of Melbourne Medical School until 1924 and played a leading role in the foundation of a school of tropical medicine at Townsville, Queensland, and of the Walter and Eliza Hall Institute, a medical research institution in Melbourne. Allen died, virtually friendless and disliked by his students and peers, in 1926.

As for Dr Camac Wilkinson, chairman of the Rabbit Commission's Experiment Committee and the man chiefly responsible for sabotaging Pasteur's bid to win the rabbit destruction prize: he returned to Australia from England in the 1890s, was an alderman with the Sydney City Council in 1902–04, then went back to London. A Harley Street ear, nose and throat specialist for many years, Wilkinson became well-known in Britain as a crusader for tuberculin. Tuberculin had been created in 1890 by Wilkinson's idol and Pasteur's rival, Germany's Robert Koch, who produced it, to be taken orally, as a remedy for tuberculosis. Koch, who died in 1910, received the Nobel Prize for Physiology/Medicine in 1906 for developing tuberculin and for his earlier bacteriological discoveries.

Camac Wilkinson became chairman of Britain's Tuberculin Dispensary League, developed 'Wilkinson's method of dosage' for tuberculin, and even wrote a book about it.[425] Unfortunately, tuberculin didn't work – either as a preventative or a cure for tuberculosis, and by the 1930s, Robert Koch's tuberculin was totally discredited. Up until Camac Wilkinson's death in London in 1948 at the age of ninety, the Australian doctor was pursued by critics who labelled him a quack and snake oil peddler for having promoted and dispensed it. In more recent times, a less spectacular use has been found for tuberculin – as an antigen that aids diagnosis of tuberculosis.

Which brings us to Brisbane, 1897. Charles Pound, the man who took the job regretfully declined by Adrien Loir, had by that time been Director of the Queensland Stock Institute for almost four years. Without fanfare, Pound had been conducting experiments to determine whether chicken-cholera existed in the Australian colonies. Dr Pound now published a paper in which he declared that, definitively, after numerous experiments over a considerable length of time, he could declare that chicken-cholera did exist in Australia.

Why did Pound decide to find out whether chicken-cholera was already in Australia? Perhaps he merely felt that it was a question that had been left unanswered for too long. Perhaps, too, he had been in contact with Adrien Loir, the man originally slated to occupy his post in Brisbane. In his 1893 letter declining the position, Loir had stated that he wished to remain in constant correspondence with whoever occupied the post. It is quite likely that Loir, via letter, put

Pound up to the chicken-cholera experiments, and Pound carried them out, to both satisfy Loir's curiosity and his own, and to answer a scientific question that had sat waiting for an answer ever since Oscar Katz's inadequate study of 1888–89.

Pound's finding caused an uproar in Queensland. Henry Tryon, the Government Entomologist and one of the two Queensland members of the Rabbit Commission of 1888–89, attacked Pound's findings – because Pound was contradicting one of the Rabbit Commission's key declarations, which was: 'The Commission cannot recommend that permission be given to disseminate broadcast through Australasia a disease which in other countries prevails in disastrous epidemics among fowls.' Tryon had been one of the commissioners to sign off on the Rabbit Commission finding that chicken-cholera did not already exist in Australia and that therefore its introduction could lead to a 'disastrous epidemic'. In 1889, this finding had sounded the death knell to Louis Pasteur's rabbit contest entry and his bid for the £25,000 prize.

Henry Tryon took Pound's findings as a personal slight, and mustered all his friends in the medical community to his support. Dr Pound was assaulted from all sides. Unable to dispute Pound's chicken-cholera research, his army of critics labelled him unprofessional, and questioned his reasons for conducting chicken-cholera experiments in the first place, implying that he had done it as a way of attacking Tryon, a professional colleague. Pound's reputation was severely dented by this episode and would lead to his being sued, unsuccessfully, several times in the coming years. These would be in unrelated cases, but an undeserved label of unprofessionalism would remain attached to him like an open sore for the rest of his career, attracting those who wanted to exploit the original wound.

Despite Tryon's personal criticism of Pound, he could not dispute Pound's research that had established chicken-cholera's presence in Australia. This being the case, its use as a means of exterminating rabbits was a viable proposition, a potentially safe means of eradicating Australia's rabbit plague. Pasteur had been right all along – chicken-cholera could have been introduced into Australia without fear of a 'disastrous' outbreak among the poultry of the continent.

EPILOGUE

IN 1897, THE YEAR that Dr Charles Pound proved chicken-cholera's existence in Australia, Adrien Loir and wife Marguerite were still in Tunis. Adrien had grown a fine handlebar moustache, and Marguerite's hair was now short and frizzy, just like Sarah Bernhardt's. By this time the couple had two sons.

A daughter was born to the Loirs in 1900. The following year, Loir upped and left Marguerite for their governess, Helen de Montes, and in 1902 he began seven years of microbiological adventures around the world with Helen at his side – in Rhodesia fighting rabies, two missions to South America, three years in Canada with the Montreal Faculty of Medicine researching horse diseases in the Rocky Mountains.

Loir's divorce from Marguerite came through in 1907 and he married Helen that same year. Loir and his second wife would have four children, one of whom, a daughter, became a doctor, married an American named Hemphill, and moved to the US. Finally, in 1909, Helen Loir was able to convince Adrien to settle down in France, and he became Director of the Office of Hygiene at Le Havre, a post totally unrelated to the Pasteur Institute or his late uncle. Loir had at last emerged from Louis Pasteur's shadow. Over the coming years, Loir held several different scientific posts in Le Havre, culminating in the directorship of the Oceanographic Laboratory at the University of Caen.

Adrien Loir died in 1941 without ever having returned to Australia since his hurried 1893 departure. He is buried in Paris's Le Pere Lachaise Cemetery, where Sarah Bernhardt was laid to rest in 1923. Following her 1891–93 world tour, which netted her 3.5 million francs,[426] Sarah did just what she had told Loir she would do while on Rodd Island: she bought a summer retreat on an island, Belle-Ile, off the Brittany coast not far from Vanne. Her retreat was an old Napoleonic fort beside the sea. 'As I never went there,' Adrien Loir later said, 'I don't know if she installed cannons, and I never thought to find out.'[427] She did.

As for the furry villains of this story: the rabbits of Australia continued to run riot, although numbers never equalled those experienced in the bumper year of 1887 when the rabbit population seemed to explode before changing climate conditions naturally culled many millions in succeeding years. In the 1950s, plague numbers swelled again and Australian authorities took a leaf out of Louis Pasteur's book and launched biological warfare against the continent's rabbits, using the myxoma virus, the cause of fatal myxomatosis. 'Myxo' as it became popularly known in Australia, reduced the rabbit population temporarily, until the bunnies developed immunity to it. The European and Spanish rabbit fleas were later introduced as disease carriers, but they could not take Australia's midsummer heat. In 1996, the RHD RCD calcivirus was introduced, with some success. But neither it nor the thousands of miles of rabbit-proof fencing built across the continent over more than a hundred years have done away with the rabbit pest.

In 2004, the Invasive Animals Cooperative Research Centre (IACRC) was established at the University of Canberra. In late 2007, the centre reported that over the past three years rabbit numbers in Australia had steadily increased. At the same time, the centre rang a climate change alarm bell, warning that the rabbit infestation was putting the carbon offset industry's investment in trees at risk – because rabbits eat tree seedlings before they can grow.

Said the centre's CEO, Dr Tony Peacock: 'Given that carbon sequestration through tree plantations and natural regeneration is firmly on the Government agenda, Australia needs to act now, before the rabbit scourge returns with a vengeance. To ensure these carbon gains are not lost, the most cost-efficient solution is to work out the biocontrol agents, such as myxoma virus and calcivirus or by looking for new agents. It is critical to do the research now rather than waiting until rabbit numbers and impacts sky rocket.'[428]

How different, we might wonder, would the situation be had Louis Pasteur won the rabbit competition prize in 1889 and his remedy been employed back then? If that had been the case, would Australia still be plagued by rabbits? Then, of course, there is the question of whether Pasteur's remedy could be successfully applied to solve the rabbit plague today. I put this question to the IACRC's CEO, Tony Peacock. His response was that he doubted whether chicken-cholera would work as a rabbit eradicator. Echoing past concerns, he didn't believe that it would spread from rabbit to rabbit sufficiently well to justify introduction. It is his belief that a long infective period is needed to ensure the disease spreads around the population.

'That's why myxo weakened as a virus,' said Dr Peacock. 'In an evolutionary sense the virus "wants" to keep the animal alive and replicating the virus's DNA. So it is not in the interest of the virus to kill the host quickly – hence ebola is a terrifying disease but burns out quickly and so is not a very "successful" virus. More importantly, the introduction of fowl-cholera would require action by the domestic chicken industry to vaccinate against the disease, and therefore I don't think the disease could be introduced due to the politics. Chicken farmers would cry foul – pun intended – and would probably win the debate – governments are risk averse, naturally, about something that might have a negative impact on an industry.'[429]

Resistant poultry farmers, politics, and governments averse to taking risks? Nothing much, it seems, has changed in 120 years. But will we ever see the likes of Pasteur, Loir or Parkes again?

APPENDIX:

Other Rabbit Competition Entries, and Dr Katz's Experiments and Recommendations

Mr West and Mr Raphael of London failed to provide any further information to support their contest submission. A follow-up letter from one of the French entrants went astray. Two other gentlemen to whom Rabbit Commission Secretary Hugh Mahon wrote could not be traced. In addition, during the course of the Commission's deliberations, the Italian consul in Sydney, Dr V. Marano, approached the Commission seeking permission to use the Rodd Island facility for two weeks to test a new biological rabbit remedy that he claimed to have discovered. After the Experiment Committee asked for a schedule of his intended experiments, Dr Marano was never heard from again.

Of the more serious entries, apart from the Pasteur scheme, the Commission explored Professor Watson's scheme, the Ellis–Butcher scheme, and a proposal from Mr Coleman Phillips of New Zealand.

Professor Archibald Watson, Australian-born Professor of Anatomy at Adelaide University, had taken his doctorate at Gottingen, Germany, before studying at the Sorbonne in Paris and also under Joseph Lister in London, so his biological credentials were excellent. An eccentric, erratic, histrionic man, he had a flowing moustache and habitually wore a canvas coat. Appearing before the Rabbit Commission at its 23 May sitting in Adelaide, he told of his experiments with rabbits suffering from a frequently fatal, naturally occurring 'rabbit scab' disease that he had been unable to identify.

By the time that Watson made his Commission appearance, he had sold all his infected rabbits to farmers who hoped to spread the disease among their properties' rabbit populations – one witness told the Commission that a farmer had paid £100 for a single rabbit suffering from Watson's disease. One such farmer was Jim Ormond, the squatter who had spotted his first rabbit west of the Murray River while travelling up the river on a paddle-steamer in 1882. 'A great many of us tried Professor Watson's scab disease,' Ormond told the Commission.[430] But he, like the others, had found the results disappointing.

Because Watson had disposed of all of his infected rabbits, the professor had nothing to show the Commission when they visited Adelaide, and could send neither live nor dead rabbits infected with this scab to Oscar Katz in Sydney for analysis or experiment. While Watson felt the disease met all the rabbit contest's requirements, he confessed that it only worked in the damper months and disappeared altogether in summer. For this reason, and the fact that no tests could be carried out on the disease even though it had been well documented in South Australia, the Commission was to say that it could not recommend any further consideration of Watson's disease.

Dr Katz had been able to study rabbits suffering with the Ellis–Butcher disease which, because it was prevalent on a 114,000-hectare section of Tintinallogy Station near Wilcannia, had come to be popularly known as Tintinallogy disease. Katz could not identify nor cultivate the disease. Neither had the Experiment Committee's Drs Wilkinson and Bancroft when they visited Tintinallogy Station and conducted postmortems on rabbits that had recently died showing Tintinallogy disease symptoms. Tintinallogy disease subsequently disappeared, and could never be identified. It may have been an epizootic such as *Coccidium oviforme*, in which parasites invade the liver and black bile ducts, although the present-day experience with this condition is that it is mostly prevalent in wet areas, and death is rare other than in young rabbits.

Many witnesses who appeared before the Commission in the winter of 1888 testified to having seen rabbits in their districts die from an affliction that had very similar symptoms to that of Tintinallogy disease. And most of them, such as Harry Vindin, Superintending Inspector with New South Wales' Rabbit Department, attributed the condition to the rabbits' habit of eating astringent tree bark when all other feed had been exhausted. These men had seen trees that had been stripped of their bark to a height of four metres (twelve feet) by rabbits, and several told of actually seeing rabbits high in trees in the arid, more remote areas. This tree bark, said Vindin and others, was indigestible, and slowly killed any rabbit that ate it.

The Experiment Committee would come to the same conclusion as Vindin, but not before attempting to secure fifty rabbits suffering from Tintinallogy disease for Oscar Katz to experiment with, and so learn more about Tintinallogy disease. Drs Ellis and Butcher demanded the payment of expense money before they would send their rabbits to Sydney. When the Experiment Committee demurred on payment, the doctors refused to part with their rabbits. The Commission was to remark that it seemed very strange that Ellis and Butcher would sacrifice their bid for the £25,000 competition prize for the sake of expense claims totalling less than £100.

Coleman Phillips, a farmer from the Wairarapa district of New Zealand's North Island, had promoted the use of *coenurus*, or bladder worm, to kill rabbits, in combination with the employment of thousands of ferrets, stoats and weasels imported from England. But there was no scientific evidence to support his entry, and no afflicted rabbits, alive or dead, available for experimentation to prove that bladder worm was an effective and contagious biological rabbit destructor. Phillips' competition entry also fell by the wayside.

In all, the Commission heard from thirty witnesses through late May and early June, and later in October 1888, at sittings in Melbourne, Adelaide and Silverton, and also at Albemarle and Tintinallogy stations in New South Wales, and lastly in Sydney. Three of those witnesses had submitted entries to the rabbit eradication contest; the remainder were squatters, station managers, rabbit inspectors and other government officials.

While the witnesses were quizzed about all aspects of the rabbit problem in their areas, the focus of inquiries was always rabbit-proof fencing. This came about because Victorian commissioner Edward Lascelles had convinced the Commission at its 26 April sitting to consider recommending to the government that it both subsidise rabbit-proof fencing and legislate that landowners must instal it to combat the rabbit plague. His argument was that, should the Commission find none of the biological competition acceptable, the only alternative was confining the rabbits with fine-wire fencing installed along all property boundaries in the infested districts, and then systematically killing them off with existing poisons.

Lascelles had two very good reasons for pushing for this option. Firstly, should any other biological method be introduced and succeed in wiping out the rabbits of Australasia, the Lascelles and Anderson rabbit poison distributor would be put out of business; and almost every squatter interviewed by the Rabbit Commission testified to having used the Lascelles and Anderson machine. Secondly, Lascelles stood to gain from the sale of rabbit-proof fencing.

Having disposed of the other competition entries, the Commission had only to deal with Pasteur's proposed method. The Commission's final recommendation regarding this was to be based on the reports of its Chief Expert Officer, Dr Katz, covering the experiments of Pasteur's representatives as well as his own experiments as prescribed by the Commission. Katz's report on the experiments by Pasteur's representatives, presented to the Commission in October 1888, had been positive regarding two aspects and inconclusive in relation to the third. In Katz's first report on his own experiments, written on 21 September of that year, he remarked that 'Australian rabbits are, to judge from a number of experiments conducted, highly susceptible to infection by the virulent microbe of chicken-cholera through their digestive organs.'[431]

On the important question of contagion, in Katz's experiments the disease definitely passed from infected to uninfected rabbits kept in wooden hutches, and killed them. Katz thought that this might also have occurred in the artificial burrows, but had no way of proving it. 'This, of course, is more or less supposition, in support of which I am unable as yet to adduce any decisive experiment. At all events I venture to suggest that a repetition of those experiments, only performed under more natural conditions – say in the open – may yield other, more favourable results.' He added, 'The possibility of successive infection in the open would also seem to be favoured by the fact that the virus of chicken-cholera preserves its infecting power in conjunction with other bacteria in putrefying or putrid substances for a considerable time.'[432]

Katz's 21 September conclusion? 'Taking everything into consideration, it would seem advisable to have further experiments conducted on a larger scale under quite natural conditions.'[433] In response to this, Wilkinson and the Experiment Committee ordered Katz to conduct 'larger scale' experiments, but not on a sheep station as Pasteur wanted or as Katz envisaged – Katz was instructed to conduct these experiments at Rodd Island.

As a consequence, in November Katz turned one hundred rabbits loose in the 'birdcage' on Rodd Island to act as 'control' animals. Contrary to the April recommendation of Dr Bancroft, the Experiment Committee continued to source wild rabbits, once more bringing them to Sydney from the Hay district. A further twenty-two of these Hay rabbits given food infected with chicken-cholera; of these, twenty died from chicken-cholera and two from starvation. Of the one hundred control rabbits, seventy-nine died from starvation. Deeming this unsatisfactory, on 19 December the Experiment Committee ordered that Katz run the experiment again. By this time, it appears, Fred Dillon Bell was on his way home to New Zealand for Christmas, so Katz found himself a new assistant, J.P. Meagher.[434] What Meagher's qualifications were or how long he worked with Katz are unknown; Bell rejoined Katz on Rodd Island, possibly as early as the following February.

But Katz had to wait for more rabbits to arrive on Rodd Island before he could commence the repeated experiment. While he was waiting, on 2 January 1889 a fire broke out on the island. The blaze destroyed the kitchen and damaged the stable as well as the connecting passages and the main enclosure. The exact cause of the fire was never determined. The enclosure was repaired by 4 February, but the kitchen was not operating again until some time later. Between 7 and 30 January, almost two hundred wild rabbits were shipped to the island to be used in the repeated experiment but, although they were well cared for, only sixty-eight were still alive when Katz began experimenting again on 12 February, at the height of summer.

The nature of this experiment was apparently conceived by Katz himself; his assistant Fred Bell had proposed the experiment to the Commission in October, and he and his fellow commissioners had approved it. This time, the enclosure was split in two. Fifteen rabbits that had been infected with chicken-cholera were introduced into one section occupied by fifty healthy rabbits. The diseased rabbits swiftly died. Five of the other rabbits caught the disease and also died; another thirty-two starved to death. In the second half of the enclosure, the 'control' section, thirty-eight rabbits were placed. Two died from starvation almost immediately. Two days later, when another shipment of rabbits arrived, the number of rabbits in the control section was increased to fifty. After three weeks, twenty-nine control rabbits were also dead from starvation. As a control experiment, this proved a dismal failure. Katz put down the high mortality rate to the fact that the test rabbits were already sickly when they reached Rodd Island, combined with the oppressive summer heat.

The Experiment Committee should have instructed Katz to repeat the experiment with healthy rabbits, but it did not. Instead, it ordered Katz to conduct several other chicken-cholera experiments. One was with ferrets, and Katz concluded that, unlike rabbits, ferrets were unaffected by chicken-cholera.

More importantly, there was a lingering question about whether chicken-cholera already existed in the colonies. It was vital to establish this – if it already existed, then the dangers of it spreading and doing harm to poultry and birds if used against rabbits were minimal, as Katz himself observed. Importantly, the ban on the introduction of imported diseases would not apply to it. Anecdotal evidence suggested that chicken-cholera had been prevalent in the Australian colonies for some years, and press advertisements placed by the Commission asking for evidence of the existence of chicken-cholera resulted in Oscar Katz being provided with nine dead chickens for analysis – in each case the owners felt sure that their birds had been killed by chicken-cholera.

In the laboratory, Katz was not able to identify chicken-cholera in any of these nine chickens. But perhaps this had more to do with Katz's lack of skill as a microbiologist. To his credit, he said himself that his unsatisfactory tests on the nine fowls did not constitute convincing evidence of the absence of chicken-cholera in Australia or New Zealand. He reported: 'In order to form a definite opinion as to whether this infectious disease exists in Australasia or not, further examinations are required and must be continued for some time.'[435]

Even though the competition rules had not mentioned birds, the Commission still required Katz to carry out experiments on indigenous birds. He found that in the laboratory chicken-cholera could kill a variety of native birds, namely the magpie, butcher-bird, blue-jay, gallah, wonga-pigeon, bronze-wing pigeon, and quail; although in the cases of the gallah, wonga-pigeon and quail, it required repeated doses of chicken-cholera to finally kill test birds. Katz concluded that chicken-cholera was much more fatal to rabbits than it was to birds.

There had never been a report of a native bird being killed by chicken-cholera in the Australian wild; if chicken-cholera did in fact already exist in Australia, this suggested that the disease did not readily transmit from infected fowls to native birds, and would not readily transmit from infected rabbits to native birds. But such a conclusion could only be reached after a broad definitive study had been carried out to prove that chicken-cholera did already exist in Australia.

On 9 March 1889, Katz summarised his final series of experiments with a question: 'Are we entitled to utilise the results of this experiment so as to draw from it definite conclusions as to the probable result of the application of chicken-cholera as a means of exterminating rabbits on a large scale? In my opinion only to a certain extent.' Katz felt that conditions in rabbit-infested areas of inland Australia would be very different from those on an island in Sydney Harbour. 'Without having the result of actual trials under the different conditions as the open country offers them, it is impossible to give a decisive answer.' His final words were: 'From the mere reason, however, that wild rabbits exhibit an extraordinary susceptibility to the microbes of chicken-cholera, and that, as I must agree with Pasteur, the susceptibility is greater in the case of rabbits than in that of birds, trials in the open, provided, of course, the arrangements can be unobjectionable, would be at once interesting and instructive.'[436]

The Rabbit Commission's Progress Report subsequently recommended that Pasteur be permitted to carry out that trial in open country, but without any involvement from the government, and purely as a courtesy to Pasteur. The Commission was saying that no matter what the outcome of such an open country trial, it would have no influence on the Commission's findings. But despite this, five of the eleven commissioners dissented against that recommendation, believing that no further trial should be permitted under any circumstances – these were Allen, Bancroft, Paterson, Stirling and Pearson.

In addition, two commissioners dissented against the Commission recommendation that the government legislate to force all landowners to erect rabbit-proof fencing – one was South Australia's Stirling; the other, Bell, felt that this was impractical in New Zealand conditions.

BIBLIOGRAPHY

Books

Angus, Beverley M., 'Pound, Charles Joseph', *Australian Dictionary of Biography*. Melbourne, MUP, 2005.

Bavin, Thomas, *Sir Henry Parkes: His Life and Work*. Sydney, Angus and Robertson, 1941.

Bernhardt, Sarah, *My Double Life*. New York, SUNY Press, 1999.

Bolton, G.C.,'Ellis, Henry Augustus', *Australian Dictionary of Biography*. Melbourne, MUP, 1981.

Bygott, Ursula M.L., *A History of the McGarvie-Smith Institute, 1918–1992*. Sydney, Sydney University, 1994.

Campbell, K.O., *Henry Parkes' Utopia*. Chetenham, NSW, Lynwood, 1994.

Chaussivert, J., and M. Blackman, editors, *Louis Pasteur and the Pasteur Institute in Australia*. Sydney, UNSW, 1988.

Creighton, C., 'Vaccination', *Encyclopaedia Britannica*. Edinburgh, 1888.

Cyclopedia of Tasmania. Hobart, Service, 1931.

Deakin, A., *The Federal Story*. Melbourne, Robertson and Mullens, 1944.

Debre, Patrice, *Louis Pasteur*. Baltimore, MD, Johns Hopkins, 1998.

Dicker, I.C., *J.C.M.: A Short Biography of James Cassius Williamson*. Rose Bay, NSW, Tudor, 1974.

Dubois, R., *Pasteur and Modern Science*. Madison, Sc Tech, 1960.

Duclaux, Emile, *Pasteur: The History of a Mind*. Philadelphia, Saunders, 1920.

Fenner, F., *A History of Microbiology in Australia*. Canberra, Austn Microbiology Society, 1990.

Fraser, C., *Come to Dazzle: Sarah Bernhardt's Australian Tour*. Sydney, Currency, 1998.

Geison, G.L., *The Private Science of Louis Pasteur*. Princeton, NJ, PUP, 1995.

Gibney, H.J.,'Mahon, Hugh', *Australian Dictionary of Biography*. Melbourne, MUP, 1986.

Hassall, J.S., *In Old Australia: Records and Reminiscences from 1794*. Brisbane, Hews, 1902.

Holmes, O.W., *One Hundred Days in Europe*. Boston, Houghton Mifflin, 1887.

Janssens, P.G., M. Wery and S. Paskoff, *Adrien Charles Loir: Pasteurien de Premier Generation*. Brussells, Academie Royal des Sciences d'Outre-Mer, 2007.

King, C.J., 'Bruce, Alexander', *Australian Dictionary of Biography*. Melbourne, MUP, 1989.

Kingsmill, A.G., 'Walker, Richard Cornelius Critchett', *Australian Dictionary of Biography*. Melbourne, MUP, 1976.

Loir, A., *La Microbiologie en Australie*, Paris, Stenheil, 1892.

Lyne, C.E., *Life of Sir Henry Parkes*. Sydney, Robertson, 1896.

Martin, A.W., *Henry Parkes: A Biography*. Melbourne, MUP, 1980.

Mitchell, A.M., 'MacLaurin, Sir Henry Normand', *Australian Dictionary of Biography*. Melbourne, MUP, 1986.

Nobel Lectures, Physiology or Medicine 1901–1921. Amsterdam, Elsevier, 1967.

Parkes, H., *Fifty Years in the Making of Australian History*. London, Longman Green, 1892.

Parsons, Vivinne, *Australian Dictionary of Biography*, 'Jamison, Thomas, (1753–1811)'. Melbourne, MUP, 1967.

Pasteur, L., *Correspondence Générale* (edited and annotated by L.P. Valéry-Radot). Paris, Flammarion, 1951.

Pasteur, L., *Germ Theory and its Application to Medicine*. Amherst, NY, Prometheus, 1996.

Pasteur, L., *Oeuvres de Pasteur* (edited by L.P. Valéry-Radot). Paris, Masson, 1922–39.

Peterson, Kamoya, 'Smith, John McGarvie', *Australian Dictionary of Biography*. Melbourne, MUP, 1988.

Reynolds, M.D., *How Pasteur Changed History: The Story of Louis Pasteur and the Pasteur Institute*. Bradenton, Florida, McGuin and McGuire, 1994.

Russell, K.F., 'Allen, Sir Harry Brookes', *Australian Dictionary of Biography*. Melbourne, MUP, 1979.

Rutledge, M., 'Burns, John Fizgerald,' *Australian Dictionary of Biography*, Vol 3. Melbourne, MUP, 1969.

Stross, R., *The Wizard of Memlo Park: How Thomas Alva Edison Invented the Modern World*. New York, Crown, 2007.

Swain, J., *The Vaccination Problem*. London, Swain, 1936.

Travers, R., *The Grand Old Man of Australian Politics: The Life and Times of Sir Henry Parkes*. Sydney, Kangaroo, 2000.

Valéry-Radot, R., *Louis Pasteur: His Life and Labours*. New York, Appleton, 1885.

Valéry-Radot, R., *The Life of Pasteur*. New York, Doubleday, 1923.

Vincent, T., '*Le Docteur Adrien Loir en Australie*', in *L. Pasteur: Savant et Société aux XIX et XX siècles* (Michel Wornonoff, ed.). Dole, Université de Franche-Compte, 1995.

Walker, W., *Recollections of Sir Henry Parkes*. Sydney, Davies, 1896.

Walsh, G.P., 'Abigail, Francis', *Australian Dictionary of Biography*. Melbourne, MUP, 1969.

Walsh, G.P., 'Lamb, John de Villiers', *Australian Dictionary of Biography*. Melbourne, MUP, 1974.

Walsh, G.P., 'Lamb, Walter', *Australian Dictionary of Biography*. Melbourne, MUP, 1974.

Webster, R., *Bygoo and Beyond*. Ardlethan, NSW, Webster, 1956.

Wessels, Sheila F., 'Lascelles, Edward Harewood', *Australian Dictionary of Biography*. Melbourne, MUP, 1974.

Wilkinson, W.C., *The Tuberculin Dispensary for the Poor*. London, Nisbet, 1923.

Journal Articles

Galland, Toby, '11 January 1887, the Day Medicine Changed: Joseph Grancher's Defense of Pasteur's Treatment of Rabies'. Baltimore, MD, *Bulletin of the History of Medicine*, Winter 2002.

Katz, Oscar, 'Bacteriological Notes on "Air-gas" for Bacteriological Work'. Sydney, *Proceedings of the Linnean Society of NSW, 1889*.

Katz, Oscar, 'Experimental Researches with the Microbes of Chicken-cholera'. Sydney, *Proceedings of the Linnean Society of NSW, 1889*.

Loir, Adrien, '*A l'ombre de Pasteur*' ('In the Shadow of Pasteur'). Paris, Le Mouvement Sanitaire, 1938.

Loir, Adrien, 'Notes on spontaneous disease among Australian rabbits'. Sydney, *Journal and Proceedings of the Royal Society of NSW*, 1891.

Loir, Adrien, 'Notes on the large death rate among Australian sheep in country infected with Cumberland disease, or Splenic fever'. Sydney, *Journal and Proceedings of the Royal Society of NSW*, 1891.

Martinez-Palomo, Adolpho, 'The Science of Louis Pasteur: A Reconsideration', Chicago, *The Quarterly Review of Biology*, Chicago University, March 2001.

Newspapers and Journals

Australasian Pastoralists' Review, 1887–98.
Brisbane *Courier*, 1888–95.
Brisbane *Evening Observer*, 1888–92.
Brisbane *Queenslander*, 1888–92.
Illustrated Sydney News, 1888–98.
La Nature, Paris, 1892 and 1893.
Launceston *Examiner*, 2007.
Le Siècle, Paris, 1887–93.
Le Temps, Paris, 1887–88.
London *Times*, 1887–1888.
Melbourne *Age*, 1888–89, 1891.
Melbourne *Argus*, 1888–89, 1891.
New York Times, 1888.
Revue de Deux-Mondes, Paris, 1887.
Revue Scientifique, Paris, 1883 and 1893.
Salut Public, Lyon, 1892.
Sydney *Daily Telegraph*, 1887–95.
Sydney *Mail*, 1888–92.
Sydney *Evening News*, 1891.
Sydney Morning Herald, 1888–95.
Sydney *Republican*, 1888.
Sydney *Tribune*, 1889.
Weekly Times, Melbourne, 1883–92 and 1927.

Official Reports

Anthrax (Cumberland) Disease in Sheep and Cattle: test of Efficacy of Pasteur's Vaccine of Anthrax as a Preventative Against – Report of Experiments at Junee, etc., NSW Legislative Assembly, December 1888.

Cumberland Disease, Correspondence Respecting Treatment of, NSW Legislative Council, 1889.

Cumberland Disease, Report of the Board on, to the Secretary for Mines, NSW, August 1888.

Katz, Dr Oscar, Report on the Biological and Microscopical Examination of Samples of Water from the Coliban Supply, Melbourne, Victorian Govt Printer, 1890.

McGarvie-Smith, J., Investigation of the Composition of the air in the Sewers of Sydney, with special reference to the presence of germs. Sydney, Govt Printer, 1894.

McGarvie-Smith, J., Mr. McGarvie-Smith's Report in the Ventilation of Sewers. Sydney, Govt Printer, 1894.

Progress Report of the Royal Commission of Inquiry into Schemes for Extermination of Rabbits in Australasia, NSW Legislative Assembly, April 1889.

Rabbit Bill, (Message No. 44), NSW Legislative Assembly, 1890.

Rabbit Nuisance Act, Return of Annual Expenditure Under, NSW Legislative Assembly, 1890.

Report of the Royal Commission Appointed to Inquire into the Introduction of Contagious Diseases Amongst Rabbits, NSW Legislative Assembly, October 1889.

The Rabbit Pest, Report of Health Board and Other Papers on Proposals to Effect Destruction of Rabbits by Means of Diseases to the Introduced, NSW Legislative Council, 1888.

Vaccination, Report for 1888, NSW Legislative Assembly, 1889.

Votes and Proceedings, NSW Parliament, Legislative Assembly, Register of Public and Private Bills, 1851–1901.

Votes and Proceedings, NSW Parliament, Legislative Council, Register of Public and Private Bills, 1851–1901.

Papers and Pamphlets

Dalton, J.C., *Political Affairs in New South Wales*, 1869.

Hotel Australia. Sydney, 1893.

Loir, A., *Pasteur's Vaccine of Anthrax in Australia*. Sydney, 1890.

Matong School History. Matong, 2005.

Parkes, Sir Henry, *Present State of Public Affairs: A speech delivered at Granville,*

6 August 1886. Sydney, Turner and Henderon, 1886.

Parkes, Sir Henry, *Speech at Temperance Hall, Saturday 20 July 1895*. Sydney, Turner and Hendersen, 1895.

Pasteur, Louis, *The Physiological Theory of Fermentation*. Comptes vendus de l'Académie des sciences, 1878.

The Linnean Society of New South Wales, 1874–1925. Sydney, Linnean Society, 1925.

Waldow, C.O., *The Present Depression in Trade, its Cause and Remedy*. Sydney, Jarrett, 1887.

Documents, Letters, Scrapbooks, Archives

'Alleged Detention and Opening of M. Pasteur's Correspondence', Colonial Secretary's Special Bundles, State Records of NSW, Western Sydney Records Centre, Kingswood Repository.

Loir, Adrien, Scrapbook Collection, Basser Library, Australian Academy of Science, Canberra.

London Census of 1881, the National Archives, London, Family History Library.

Middlesex County Cricket Club Archive, UK.

New South Wales Parliamentary Archives, Registry of Former Members, Sydney.

Papers of Sir Henry Parkes, Mitchell Library, State Library of New South Wales, Sydney.

Pasteur, L., Unpublished Letters, Archives de l'Ecole Normale Supérieure, Paris.

Pasteur's Vaccine Lymph, Return Respecting, Legislative Assembly, NSW, December 1891.

Press Cuttings on Sir Henry Parkes, 1880–96, Dixon Reading Room, State Library of New South Wales, Sydney.

Queensland State Archives, Brisbane, Colonial Secretaries' Correspondence.

Smithsonian Institution Archives, Washington DC.

NOTES

CHAPTER ONE – The Killing Team Arrives

1. Melbourne *Age*, 2 April 1888.
2. Ibid.
3. Loir, *La Microbiologie en Australie.*

CHAPTER TWO – Pasteur's Beer of Revenge

4. Loir, '*A l'ombre de Pasteur*'.
5. Pasteur, *Correspondence générale*, 2:492.
6. Pasteur, *The Physiological Theory of Fermentation.*
7. Debre, *Pasteur.*

CHAPTER THREE – Becoming Pasteur's Protégé

8. Loir, '*A l'ombre*'.
9. Ibid.
10. Dubois, *Pasteur and Modern Science.*
11. Duclaux, *History of a Mind.*
12. Ibid.
13. Pasteur, *Correspondence générale*, 3:159.
14. Valéry-Radot, *Vie de Pasteur.*
15. Ibid.
16. Pasteur, *Germ Theory.*
17. Debre, *Pasteur.*
18. Loir, '*A l'ombre*'; and Geison, *Private Science.*
19. Duclaux, *History of a Mind.*
20. Ibid.
21. Pasteur, *Germ Theory.*
22. Duclaux, *History of a Mind.*
23. Valéry-Radot, *Louis Pasteur, His Life and Labours.*
24. Ibid.
25. Loir, '*A l'ombre*'.
26. Geison, *Private Science.*
27. Pasteur, 'The Anthrax Vaccination,' *Revue Scientifique*, 20 January 1883.
28. Loir, '*A l'ombre*'.
29. Debre, *Pasteur.*
30. Loir, '*A l'ombre*'.
31. Ibid.
32. i. Valéry-Radot, *Louis Pasteur, His Life and Labours.* ii. Because Pasteur found that the success of his swine fever vaccine varied enormously depending on the breed of pig, he turned his back on this field of research. In 1886, French veterinarians used Pasteur's research as the basis for a vaccine with which they successfully vaccinated 100,000 pigs – Debre, *Pasteur.*

33. Pasteur, 'Anthrax Vaccination', *Revue Scientifique.*
34. Ibid.
35. Ibid.
36. Ibid.

CHAPTER FOUR – The Rabbit Invasion

37. Rabbit Commission (RC) Progress Report, Proceedings.
38. Hassall, *In Old Australia.*
39. Attributed. No original source identified.
40. RC, Proceedings.
41. Ibid.
42. *Weekly Times*, 23 July 1927.
43. RC, Proceedings.
44. Ibid.
45. *Weekly Times*, 23 April 1881.
46. *Weekly Times*, 13 September 1884.
47. RC, Proceedings.
48. *Weekly Times*, 13 September 1884.
49. RC, Proceedings. The Australian colonies were using British pounds shillings and pence as their currency. There were twelve pence to the shilling, and twenty shillings to the pound – a total of 240 pence to the pound. A 'guinea' was one pound and one shilling.
50. RC, Proceedings.
51. Ibid.
52. Ibid.
53. Ibid.
54. Ibid.
55. Ibid.
56. Ibid.
57. Ibid.
58. Ibid.
59. Ibid.
60. Ibid.
61. NSW Legislative Assembly, Votes and Proceedings, 1887.

CHAPTER FIVE – Sir Henry Parkes' Rabbit Eradication Competition

62. *Illustrated Sydney News*, 28 March 1891.
63. Lyne, *Sir Henry Parkes.*
64. Ibid.
65. Walker, *Recollections of Sir Henry Parkes.*
66. Bavin, *Sir Henry Parkes: His Life and Work.*
67. Lyne, *Parkes.*
68. Parkes, *Fifty Years in the Making of Australian History.*
69. Lyne, *Parkes.*
70. Walker, *Recollections.*
71. Ibid.
72. Ibid.
73. Sydney *Republican*, 4 April 1888.
74. Waldow, *The Present Depression in Trade.*
75. Lyne, *Parkes.*
76. Ibid.
77. Ibid.
78. *Illustrated Sydney News*, 1 October 1891.
79. Vincent, in *Dr Adrien Loir en Australie*, gives Abigail the credit for putting forward the idea.
80. Rabbit Nuisance Act, Return of Expenditure Under, NSW Legislative Assembly, 1890.

CHAPTER SIX – Paying for Pasteur's Institute

81. Debre, *Pasteur.*
82. Valéry-Radot, *Life and Labours.*
83. Loir, '*A l'ombre*'.
84. Holmes, *One Hundred Days in Europe.*
85. Reynolds, *How Pasteur Changed History.*
86. Pasteur, 'Maladies virulentes, virus-vaccins et prophylaxie de la rage', *Oeuvres de Pasteur.*
87. Ibid.
88. Debre, *Pasteur.*
89. Pasteur to Jules Vercel, *Correspondence générale*, 4:65.
90. Figure provided by Pasteur Museum, Paris, May 2007.
91. Valéry-Radot, *Vie de Pasteur.*

CHAPTER SEVEN – The Rabbit Eradication Prize

92. Duclaux.
93. RC, Final Report, 1889; Loir, *Microbiologie en Australie*. In the version printed in the Rabbit Commission report the last paragraph is omitted, but it is included in Loir.
94. Loir, '*A l'ombre*'.
95. *Le Temps*, 29 November 1887.
96. Buckley to Bell, 31 August 1886, National Archives, Wellington.
97. *Revue des Deux-Mondes*, 15 August 1887. Quoted extensively by Pasteur in 5 January 1888 letter to NSW Agent-General. Also quoted in The Rabbit Pest: Report of Board of Health and Other Papers.
98. *Le Temps*, 29 November 1887.
99. Ibid.
100. Ibid.
101. Ibid.
102. These experiments are described by Pasteur in correspondence included in the RC Progress Report.
103. Ibid. Madame Pommery's name was Jeanne Alexandrine Louise Pommery, but the translated versions of these letters to Pasteur have her inexplicably signing herself as 'Eve Pommery'. Pommery Champagne management today believes this to have been simply an error in transcription at the time.
104. Ibid.
105. *La Nature*, 1892.
106. Article extracts reprinted in RC Progress Report, Appendix III.
107. Mme Pommery to Pasteur, 3 December 1887. Reprinted in RC Progress Report, App III.
108. *New York Times*, 12 February 1888.
109. Mme Pommery to Pasteur, c. 4 January 1888 – received by Pasteur in Paris on 5 January. Referred to by Pasteur in RC Progress Report, App III.
110. Louis Momont was the nephew of Roux's stepfather.
111. Pasteur to Samuel, 5 January 1888, RC Progress Report, App III.
112. Ibid. The Royal Society of New South Wales confirms that Paste[illegible] was a member of the society at [illegible] time. In addition, they hold a [illegible] from Pasteur thanking the soc[illegible] granting him membership.
113. Pasteur to Abigail, 8 January [illegible]C Progress Report, App III.
114. The Rabbit Pest.
115. The cost of the [illegible]s from Pasteur to Australia [illegible] this period was noted on th[illegible]grams and the copies held in t[illegible]eged Detention' file. Pasteur w[illegible] soon be asking the NSW Gov[illegible]ent to cover the cost of his tele[illegible]ms to and from Loir and Germon[illegible]
116. Loir, '*A l'ombre*', and *La Microbiologie*.

CHAPTER EIGHT – Bureaucratic Warfare

117. Waldow, *The Present Depression*.
118. The Rabbit Pest.
119. Ibid.
120. Ibid.
121. Ibid.
122. Ibid.
123. Ibid.
124. Ibid.
125. Ibid.
126. *Correspondence générale*.
127. Melbourne *Age*, 3 May 1888.
128. Pasteur to Loir, 1 March 1888, Bibliothèque Nationale.
129. Melbourne *Age*, 3 May 1888.
130. The Rabbit Pest.

CHAPTER NINE – The Brewer's Offer

131. Loir, '*A l'ombre*'.
132. Ibid.
133. Ibid.
134. Ibid.
135. Carlton and United Breweries figures, June 2007.

CHAPTER TEN – The Rabbit Commission Meets

136. Neither the Rabbit Commission reports nor the press of the day stated precisely where at the Chief Secretary's Department the Rabbit Commission held its meetings. However, in discussions with the author in November 2007, Her Excellency the Governor of New South Wales, Professor Marie Bashir and her Chief of Staff, Brian L. Davies, were of the view that these meetings must have taken place in the Executive Council Chamber. There are two large meeting tables in the chamber, one used by the Governor of the day and the members of the Executive Council; the other, a committee table at the eastern, Macquarie Street, end of the room, would have been that used by the Rabbit Commission.
137. RC Progress Report, Proceedings.
138. Griffith to Gordon, 8 March 1888, Queensland State Archives.
139. RC Progress Report, Proceedings.
140. Ibid.
141. *Illustrated Sydney News*, 15 March 1888.
142. RC Progress Report, Proceedings.
143. Ibid.
144. Ibid.
145. Creighton, *Encyclopaedia Britannica*, 1888.
146. NSW Legislative Assembly, Votes and Proceedings, 1889.
147. NSW Legislative Assembly, Vaccination Report for 1888, 1891.
148. Creighton, *Encyclopaedia Britannica*, 1888.
149. Ibid.
150. RC Progress Report, Proceedings.
151. Ibid.
152. Ibid.
153. Ibid.
154. Ibid.
155. Ibid.
156. Ibid.
157. NSW Parliamentary Archives.
158. Ibid.
159. Loir, '*A l'ombre*'.
160. RC Progress Report, Proceedings.
161. Ibid.

CHAPTER ELEVEN – Fifteen Hundred Entries

162. RC Progress Report, Proceedings. Raphael was still living at Wellington Mansions, Oxford Street, London, in 1894, when he and his wife announced the birth of a son. In 1902, West, Raphael and a third partner registered a patent in Canada for a lens manufacturing machine, with West credited as its inventor. Raphael, it seems, was a financier of West's schemes.

CHAPTER TWELVE – Lost in Translation

163. Loir, '*A l'ombre*'.
164. Ibid.
165. Ibid.
166. Ibid.
167. Lascelles, before Rabbit Commission, 26 April 1888. Progress Report, Proceedings.
168. RC, Progress Report, Proceedings.
169. Melbourne *Age*, 26 March 1888.
170. Ibid.
171. RC Progress Report, Proceedings.
172. Ibid.
173. Ibid.
174. Ibid. – this entire exchange.
175. Ibid. – this entire exchange.

CHAPTER THIRTEEN – Enter Doctor Katz

176. Dr Joc Forsythe to the author, letter of 6 May 2007.
177. RC Progress Report, Proceedings.
178. Melbourne *Age*, 26 April 1888.
179. In February 1889, MacLaurin was made a member of the Legislative Council by the brief Protectionist Government of Sir George Dibbs.

MacLaurin resigned all other Government appointments, including that of Government Medical Adviser and President of the Board of Health, but subsequently spent most of 1889 on an extended private visit to Europe.

180. Walker, *Recollections*.
181. RC Progress Report, Proceedings.
182. Ibid. – this entire exchange.
183. Australian Bureau of Meteorology.
184. RC Progress Report, Proceedings.
185. Ibid. – this entire exchange.
186. Herring to Loir, 24 April 1888, Mines Department ref. 88/4115.
187. Loir, '*A l'ombre*'.
188. Ibid.
189. Ibid.
190. Ibid.

CHAPTER FOURTEEN – The Impasse

191. Based on what Salomons had told the press in London, as reported by Melbourne *Age* of 3 May 1888.
192. *Sydney Morning Herald*, 28 April 1888.
193. *Sydney Morning Herald*, 3 May 1888.
194. Loir, Cumberland Disease pamphlet, and numerous Sydney, Melbourne and Brisbane newspapers of May 1888.
195. RC Progress Report, Proceedings.
196. Ibid.
197. Ibid.
198. RC Progress Report, Section VII.

CHAPTER FIFTEEN – Robert Koch's Sabotage

199. Loir, '*A l'ombre*'.
200. This information was contained in a 1 April 1888 letter by Germont and Loir to the editor of the *Sydney Morning Herald* which was published by the paper the following day.
201. In the 1890s the French Consulate moved to Bond Street, where a French bank, the Alliance Française, and other French organisations had offices. Loir's map is today held by the Basser Library, Australian Academy of Science, Canberra.
202. *Matong School History*.
203. A report in the *Sydney Morning Herald* of 9 August 1888 details this trip by Loir to the Sydney laboratory, and his work there with Hamlet.
204. Pasteur to Lytton, 11 June 1888, Cumberland Disease, Correspondence Respecting Treatment of.
205. Ibid.
206. Lytton to Salisbury, 12 June 1888. Ibid.
207. Fergusson to Under Secretary of State, 19 June 1888, Cumberlnad Disease, Correspondence.
208. Knutsford to Carrington, 26 June 1888, Cumberland Disease, Correspondence.
209. RC, Proceedings.
210. Ibid.
211. Ibid.
212. Pasteur to Mrs Priestley, 17 June 1888, Bibliothèque Nationale.
213. Pasteur to Salomons, 18 June 1888, Bibliothèque Nationale.
214. RC, Section VI.
215. Ibid.
216. Ibid.
217. These words were repeated in Sir Daniel Cooper's 19 June 1888 telegram to Parkes. Alleged Detention file.
218. Ibid.
219. RC, Proceedings.
220. RC, Wilkinson's Experiment Committee Report of July 3, Section VII.
221. Ibid.
222. RC, Progress Report, preface.
223. RC, Proceedings.
224. Ibid.
225. There is no record of this outgoing telegram, as there is no record of any of the telegrams sent from Loir and Germont to Pasteur. The telegram's existence and likely wording can be deduced from subsequent events.
226. Parkes to Cooper, 22 June 1888. Alleged Detention file.
227. Ibid.
228. Ibid.

229. Loir, '*A l'ombre*'.
230. Ibid.
231. This is revealed by William Trail in the Legislative Assembly, 29 April 1889. Votes and Proceedings.
232. RC, Proceedings.
233. The RC Progress Report says that Fischer arrived in Sydney and gave the microbes to Katz. Katz himself, in Experimental Researches with the Microbes of Chicken-cholera, Linnean Society Proceedings, 1889, reveals the date and the fact that Fischer had brought the microbes from Koch's laboratory in Berlin.

CHAPTER SIXTEEN – Sir Henry's Covert Action

234. Pasteur to Germont, Alleged Detention Colonial Secretary's Ref. 88-7948.
235. Details of the actions of Records Clerk and Chief Clerk are revealed in Legislative Assembly debate 11 April 1889. Votes and Proceedings.
236. Contents of note revealed in *Sydney Morning Herald*, 12 March 1889.
237. Pasteur to Germont. Alleged Detention.
238. Ibid.

CHAPTER SEVENTEEN – The Rodd Island Experiments

239. Loir, '*A l'ombre*'.
240. All these details are contained in RC, Section VIII, and Katz, Experimental Researches with the Microbes of Chicken-cholera, *Linnean Society Proceedings*, 1889.
241. Katz, On Air-Gas for Bacteriological Work, *Linnean Society Proceedings*, 1889.
242. Loir, '*A l'ombre*'.
243. Katz, Experimental Researches; and Pearson, RC, Section VIII.
244. RC, Proceedings, 18 August 1888.
245. Pasteur to Loir, 7 September 1888, Bibliothèque Nationale.
246. RC, Proceedings.
247. Dr John Creed, Legislative Council debate 11 April 1889.
248. Jean Chaussivert, in *Louis Pasteur and the Pasteur Institute in Australia*, notes that Loir's daughter Marie retained calendars on which her grandmother, Loir's mother, Amelie Loir, had marked the dates she received each letter from her son while he was in Australia. Over a period of two years, Loir wrote his mother 82 letters – roughly one every two weeks while he was in New South Wales at that time.
249. King, *Australian Dictionary of Biography*.
250. Loir, *La Microbiologie en Australie.*
251. Katz, Experimental Researches.
252. RC, Proceedings.
253. Katz, RC, Section X.
254. Ibid.
255. Ibid.
256. RC, Proceedings, 18 August 1888.
257. Ibid.
258. Katz, Experimental Researches.
259. RC, Proceedings, 18 August 1888.
260. Katz, Experimental Researches. Fischer's primary interest was botany, not microbiology. There was no reason for him to suddenly turn up in Sydney with chicken-cholera microbes in his baggage other than as a result of a request from Wilkinson or Katz to bring them from Koch in Berlin to help undermine the Pasteur bid for the rabbit prize.
261. RC, Proceedings, 18 August 1888.

CHAPTER EIGHTEEN – Showdown With Sir Henry

262. Loir, *La Microbiologie.*
263. Pasteur to Germont, Alleged Detention, Ref. 88/9037.
264. Pasteur to Germont, Alleged Detention, Ref. 88/9335.
265. Ibid.
266. Legislative Assembly debate, 11 April 1889. Votes and Proceedings.
267. Ibid.
268. Ibid.

269. Germont and Loir to Editor, *Sydney Morning Herald*, 1 April 1889, published 2 April 1889.
270. Parkes, Legislative Assembly, 11 April 1889. Votes and Proceedings.
271. Germont and Loir, *Sydney Morning Herald*, 2 April 1889.
272. Legislative Assembly debate, 20 September 1888. Votes and Proceedings.
273. Germont and Loir, *Sydney Morning Herald*, 2 April 1889.

CHAPTER NINETEEN – Joy at Junee

274 Loir, *Pasteur's Vaccine of Anthrax in Australia*.
275. Pasteur to Germont, Alleged Detention, Ref. 88/10841.
276. Pasteur to Bruce, 1 October 1888, RC, Progress Report, preface.
277. Ibid.
278. Loir, '*A l'ombre*'.
279. Ibid.
280. Ibid.
281. *Matong School History*.
282. Anthrax (Cumberland) Disease in Sheep and Cattle.
283. Pasteur to Germont, Alleged Detention, Ref. 88/10869.

CHAPTER TWENTY – Pasteur's Grand Opening

284. Debre, *Pasteur*.
285. *Sydney Morning Herald* – reprinted by Brisbane *Evening Observer*, 1 January 1889.
286. Ibid.
287. Figures supplied by Pasteur Museum, Paris, April 2007.
288. Author's estimation.
289. Debre, *Pasteur*
290. *Sydney Morning Herald* – reprinted in Brisbane *Evening Observer*, 1 January 1889.
291. Pasteur, *Mélanges scientifiques et littéraires.*
292. *Sydney Morning Herald* – reprinted in Brisbane *Evening Observer*, 1 January 1889.

CHAPTER TWENTY-ONE – Traitors in the Camp

293. Brisbane *Courier*, 13 December 1888.
294. Loir, '*A l'ombre*'.
295. Brisbane *Evening Observer*, 26 March 1889.
296. Loir, '*A l'ombre*'.
297. Ibid.
298. Brisbane *Evening Observer*, March 1889.
299 Parkes, *Fifty Years*.
300. Lyne, *Sir Henry Parkes*.
301. Parkes, *Fifty Years*.
302. Pasteur to Germont, Alleged Detention, Ref. 89/2944.
303. Martin, *Henry Parkes*.
304. Sydney *Truth*, March 15 1891.
305. Lyne, *Sir Henry Parkes*.
306. Ibid.
307. Brisbane *Courier*, March 1889.
308. *Sydney Morning Herald*, December 1888, with a Melbourne dateline of 21 December.

CHAPTER TWENTY-TWO – The Progress Report

309. Brisbane *Courier*, 6 April 1889.
310. Sydney *Daily Telegraph*, 3 April 1889.
311. RC, Progress Report, preface.
312. Ibid.
313. Ibid.
314. Ibid.
315. Ibid.
316. Ibid. Note the wording 'Dr Katz indicated', not 'Dr Katz stated'. This appears to be deliberate, as if to excuse the distortion by the report's authors.
317. Ibid.
318. Ibid.
319. Ibid.
320. Ibid.
321. Ibid.
322. Ibid.
323. RC, Proceedings.
324. Ibid.
325. Loir, *La Microbiologie*.

CHAPTER TWENTY-THREE – An International Incident

326. *Sydney Morning Herald*, 12 March 1889.
327. *Sydney Morning Herald*, 21 March 1889.
328. *Sydney Morning Herald*, 22 March 1889.
329. Sydney *Daily Telegraph*, 30 March 1889.
330. *Sydney Morning Herald*, 2 April 1889.
331. 29 March, 1889
332. 23 March, 1889
333. *Bulletin*, March 1889.
334. Parkes, Legislative Assembly, Answer to Question No. 2, 3 April 1889.Votes and Proceedings.
335. Legislative Council Debate, 3 April 1889,Votes and Proceedings.
336. Legislative Assembly debate, 11 April 1889,Votes and Proceedings.
337. Ibid.
338. Pasteur added these observations to a letter which his wife Marie wrote to Eliza Priestley on 2 May 1889 about the Rabbit Commission's Progress Report. Bibliothèque Nationale.

CHAPTER TWENTY-FOUR – Back to Sydney

339. *La Sièclе*, Paris, 12 March 1890.
340. Ibid.
341. Ibid.
342. NSW Rabbit Bill, 1890.
343. Loir, *La Microbiologie.*
344. Loir, '*A l'ombre*'.
345. *Illustrated Sydney News*, 23 May 1891.
346. Ibid.
347. Fraser, *Come to Dazzle.*
348. Ibid.
349. Loir, '*A l'ombre*'.
350. Fraser, *Come to Dazzle.*
351. Loir, '*A l'ombre*'.
352. Ibid.
353. Ibid. A story would gain currency that one of Bernhardt's dogs was pregnant and would give birth to a litter of seven pups on Rodd Island. Loir, as guardian of the two dogs, would have been aware of this, had it occurred, but he made no reference to such an event in his otherwise very detailed account of his relations with Bernhardt during her time in Sydney. It would appear that the pregnant dog story, like several other claims regarding Bernhardt's activities in Sydney including supposed dinner engagements and country excursions, were colourful fabrications by star-struck locals.

CHAPTER TWENTY-FIVE – The Sarah Bernhardt Affair

354. Fraser, *Come to Dazzle.*
355. Bernhardt, *My Double Life.*
356. Fraser, *Come to Dazzle.*
357. Loir, '*A l'ombre*'.
358. Fraser, *Come to Dazzle.*
359. Loir, '*A l'ombre*'.
360. Debre, *Pasteur.*
361. Fraser, *Come to Dazzle.*
362. Loir, '*A l'ombre*'.
363. Ibid.
364. Ibid.
365. Ibid.
366. Ibid.
367. Fraser, *Come to Dazzle.*
368. Loir, '*A l'ombre*'.
369. Ibid.
370. Ibid.
371. Ibid.
372. Ibid.
373. Ibid.
374. Fraser, *Come to Dazzle.*
375. Ibid.
376. Loir, '*A l'ombre*'.
377. Ibid.

CHAPTER TWENTY-SIX – Dr and Mrs Loir

378. *La Sièclе*, Paris, 6 February 1892.
379. Ibid.
380. *Salut Public*, Lyon, 24 July 1892.
381. Chaussivert, *Louis Pasteur and the Pasteur Institute of Australia.*

382. Bibliothèque Nationale.
383. Chaussivert, *Louis Pasteur.*
384. Loir, '*A l'ombre*'.
385. Francoise Michel-Loir to the author, 10 July 2007.
386. Ibid.

CHAPTER TWENTY-SEVEN – A Change in the Tide

387. Brisbane *Queenslander*, 18 August 1892.
388. *Australasian Pastoralists' Review*, Sydney and Melbourne, 15 September 1892.
389. Reprinted in *Australasian Pastoralists' Review*, Sydney and Melbourne, 15 March 1892.
390. Sydney *Daily Telegraph*, September 1892.
391. Loir scrapbook, September 1892.
392. Ibid.
393. *Australasian Pastoralists' Review*, Sydney and Melbourne, 15 September 1892.
394. Ibid.
395. Travers, *Grand Old Man.*
396. *Illustrated Sydney News*, 18 August 1892.
397. Travers, *Grand Old Man.*
398. Bruce to Loir, 7 January 1892.
399. Brisbane *Evening Observer*, 19 September 1892.
400. Brisbane *Evening Observer*, 16 September 1892.
401. Loir scrapbook, 1892.
402. Sydney *Daily Telegraph*, 10 October 1892.
403. Sydney *Daily Telegraph*, 12 October 1892.
404. Ibid.
405. *Sydney Morning Herald*, 13 October 1892.
406. *Illustrated Sydney News*, 1 October 1892.
407. Ibid.
408. Sydney *Daily Telegraph*, 20 September 1892.
409. Reprinted in the Sydney *Daily Telegraph*, 20 September 1892.
410. Ibid.
411. Ibid.
412. Chausssivert, *Louis Pasteur.*
413. *La Siècle*, Paris, 12 March, 1890.
414. Loir, '*A l'ombre*'.
415. *Australasian Pastoralists' Review*, Sydney and Melbourne, 15 March 1893.
416. Francoise Michel-Loir to the author, 10 July 2007.
417. Ibid.
418. Janssens, Wery and Paskoff, *Adrien Charles Loir.*

CHAPTER TWENTY-EIGHT – The Final Twists

419. In Melbourne, Katz had been employed by Professor Harry Brookes Allen's royal commission into the sanitary state of Melbourne, to investigate a claim by Alphonse de Bavay, a chemist working for a Melbourne brewery, that he had isolated typhoid bacilli in Melbourne's drinking water. De Bavay gave Katz two samples of what he said were cultures taken from Melbourne's water. Katz was unable to identify typhoid bacilli in either sample. De Bavay subsequently revealed that, encouraged by Dr John Springthorpe, an iconoclastic Melbourne University lecturer and Alfred Hospital physician who wanted to embarrass the pompous Professor Allen, one of the cultures which Katz had been unable to identify had indeed been typhoid, taken by De Bavay from a victim of clinical typhoid. Allen's reputation was left undamaged, but Katz's reputation as a bacteriologist took a direct hit, even though he would protest that Robert Koch had previously pointed out how difficult it then was to isolate typhoid bacilli from water. (Details of the De Bavay affair provided to the author by Dr Joc Forsythe in a letter of 6 May 2007.)
420. Lyne, *Sir Henry Parkes.*
421. Ibid.
422. Webster, *Bygoo and Beyond.*

423. Bygott, *History of McGarvie-Smith Institute.*
424. Peterson, 'Smith, John McGarvie'; and Bygott, *History.*
425. Wilkinson, *Tuberculin Dispensary for the Poor.*

EPILOGUE

426. Dicker, *J.C.W., a Short Biography of James Cassius Williamson.*
427. Loir, '*A l'ombre*'.
428. Launceston *Examiner*, 7 October 2007
429. Dr Tony Peacock to the author, in an email of 18 October 2007.

APPENDIX

430. RC, Progress Report, Proceedings, Section II – Evidence given 15 October 1888.
431. Katz, RC, Progress Report, Section XI.
432. Ibid.
433. Ibid.
434. Katz, Experimental Researches, *Linnean Society Proceedings*, 1889.
435. Katz, RC, Progress Report, Section XII.
436. Ibid.

INDEX

Also by Stephen Dando-Collins

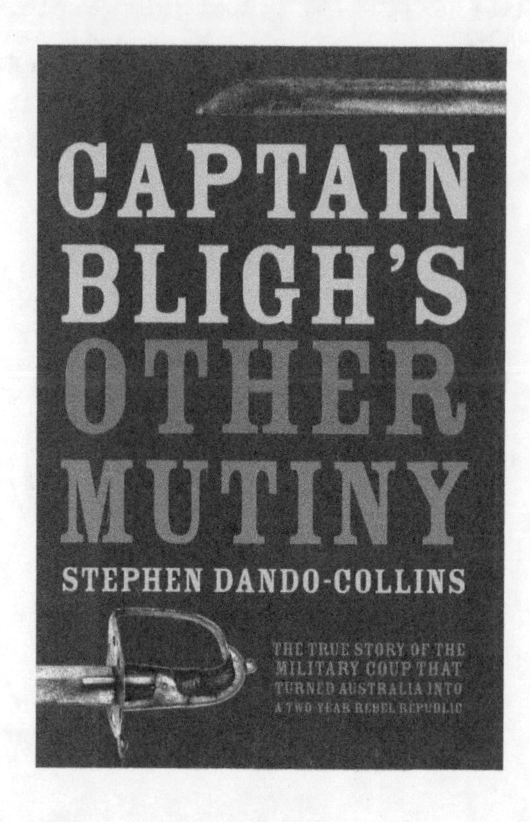